DIMENSIONS OF THE SPIRIT

'Your writings are wonderful and I have been helped and inspired by them. It is a remarkable and unique book. You have a wonderful personal story to tell... It will help and inspire others.'

Sir Colin Humphreys, Cambridge University Professor.

Everyone will be interested in these tales... I love this, Eric. You have the makings of a masterpiece here..'

Leonard Sweet, Professor of Evangelism. Drew University Theological School (Madison, NJ.) and Visiting Distinguished Professor at George Fox University (Portland, OR)

'What a lovely set of stories and images, conveying the power and thrust of the Holy Spirit's work in the world today!'

Paul Anderson, Professor of Biblical and Quaker Studies, George Fox University, Newberg, Oregon.

'Fascinating and wonderful testimonies with their scientific parallels.'

Gordon Clarke, Emeritus Director of Christian Fellowship Ministries, Stockton on Tees.

'I am interested in the science/spirituality crossover... I look forward to seeing the book in print. It will certainly cause many people to think again, and in a more positive way...'

Professor Keith Ward (who knew the author as a scientist and principal in the Science and Religion Forum), Regius Professor Emeritus at Oxford.

'It bears all your hallmarks – in particular your easy movement between the realms of the Spirit and those of science, which nobody else, in my experience, quite manages! I shall await the book with much interest...'

Mark Harrison, Head of Theology and Philosophy, Bede Sixth Form College, Stockton.

'I always think of your presentation as one of the highlights of my undergraduate career. A presentation on Faith healing in Nicholson Hall at the University of Minnesota, what could be better?'

Josiah Lindstrom, Chairman of the 'Mars Hill' group of students

DIMENSIONS OF THE SPIRIT

SCIENCE AND THE WORK OF THE HOLY SPIRIT

Eric William Middleton

First published in Great Britain by Piquant Editions in 2012. Piquant Editions is an imprint of Piquant.

www.piquanteditions.com

ISBN 978-1-903689-89-9

British Library Cataloguing in Publication Data

Middleton, Eric.
Dimensions of the spirit : the new Canterbury tales.
1. Spiritual life--Christianity. 2. Healing--Religious aspects--Christianity. 3. Holy Spirit. 4. Religion and science. 5. Physics--Religious aspects--Christianity.
I. Title
248-dc23

ISBN-13: 9781903689899

Unless otherwise stated, Scripture quotations are the author's own translations.

Cover design by Luz Design, www.projectluz.com

Typesetting by ToaTee, www.2at.com

CONTENTS

INTRODUCTION

Meeting Hurting Pilgrims: An Exciting Journey

It was the second time William had flown across the Atlantic. On his first trip, on the way to read a paper on 'The Origin of Five Dimensions' to an international physics conference in Pittsburgh, he had met the Holy Spirit.

This year he was due to give lectures and seminars at universities in North America on the science/faith dialogue. He talked about the great pillars of science in the modern age, where we find awe and wonder, mystery, and paradoxes which seem almost spiritual. Towards the end he changed gear ('morphed', as his professor host described it) to talk about his experiences of healing. He was asked to describe his spiritual journey, the stepping stones he had followed from the healing of memories and hurts from the past to physical healing—even to the deliverance and healing of places.

While on this trip Paul the professor gave William the vision of a cathedral. It could have been Canterbury, but it also resembled Durham Cathedral, near William's home. Great stone pillars of discoveries in science seemed to be marching onwards, towards the high altar. In parallel with these were pillars of spirituality and healing. Above were the double arches of mind-blowing ten or eleven dimensions, and also the overriding power of the Holy Spirit. Windows of dreams, visions and pictures illuminated the whole building. The foundations were Celtic spirituality, resting on the solid rock of Jesus.

As he travelled around the USA it was William's privilege to meet many pilgrims to this spiritual cathedral. An unusual number of fellow passengers on the various flights had served in Iraq or Afghanistan, and they shared with him some of their hidden traumas and flashbacks. Almost everyone harboured hurts from the past and William was able to share the reality of the Spirit. The challenge was to find a language in which to talk about such things, and even to discuss spiritual warfare. Some had physical injuries as well as hurtful memories. These led William to share stories of healings in his talks, and some of his hearers experiencing their own inner healing even prayed with others for physical healing.

On the way William sought to demolish the perception that science and faith are in conflict—an idea frequently expressed by his barber. He has found that when we become comfortable with the paradoxes and mysteries of postmodern science, the twin pillars of science and faith actually reinforce one another under the same roof.

'This must be written in a book,' his US audience insisted. *'We need to know more!'*

William's Journey: Never Quite Knowing Where It Would Lead

William's journey in the Spirit so far has been a constant surprise and source of wonder for him and for those who have joined him along the way, involving inner healing and the healing of hurtful memories, trauma, voices, physical complaints and even of places and words said against them. These experiences triggered his interest in key breakthroughs to unifying forces and particles at the sharp end of contemporary science. He has been able to explore a world view encompassing extra dimensions of space-time alongside the dimensions of the Spirit. For William and those who travelled with him it became a pilgrimage led by the Spirit.

The spiritual dimension seemed to provide the space for a personal encounter with the Holy Spirit, the Spirit of Jesus. William was also to encounter dark spirits, almost demonic, as can be seen from the accounts told in this book. It is when he and those with him recognized the dark powers that they really knew the power of the Holy Spirit. 'After the healing touch, we are never the same again,' Richard affirmed after receiving prayer; 'inner healing has changed my life!'

Narrative Theology: Real Tales of Healing

William's story allows us to be comfortable not just with the reality of the presence of the Holy Spirit but also with new insights from postmodern science. Each chapter of this book presents aspects of contemporary physics leading to the 'theory of everything' in many dimensions. The theology of M-theory is seen as a springboard, preparing the way for meeting the dimensions of the Spirit, perhaps touching our own spiritual need for healing. In parallel a series of real stories of life-changing encounters and experiences of healing in the Holy Spirit is given. These examples bring the book to life, bridging the gap between the theoretical and experiential; the theology is not deduced but actually encountered. The stories are all given with the permission of those involved, and the whole journey can be seen as 'narrative theology' rather than academic theology: healing in practice in the name of Jesus.* Often the names of individuals have been changed because they shared the more intimate details of their lives.

This book takes seriously both the postmodern view of reality and also the Holy Spirit from within Christian theology. It is a springboard of personal testimonies of healing and wholeness to emphasize a new world view of reality: the multidimensions of M-theory *and* the dimensions of the Spirit. As a scientist William often says, 'What if we try ...?' You are invited to come on a journey with William and to follow these extra dimensions of the Spirit's appearing—a true voyage of discovery!

* Narrative theology is a paradigm shift from the systematic theology of the past: from Judy Naegli, quoted in Danielle Shroyer, *The Boundary-Breaking God: An Unfolding Story of Hope and Promise* (San Francisco: Jossey-Bass, 2009).

Chapter 1

THE ARTIST'S TALE: STEPPING STONES TO A DEEPER HEALING

It all started with Josh. 'Can you get rid of the voices and dark spirits which are haunting me?' he challenged William. Josh was a very artistic sixth-form student at a college in the North of England, with intense blue eyes, black eyebrows and ruffled hair. He had fooled around with spirit guides and was very involved in Heavy Death Metal rock music. He had also accepted a friend's invitation to a spiritualist 'church'. After attending there he had become aware of voices and powerful dark spirits, and he was very disturbed about it. Eventually he came to see William, the principal of the college, in his rooms and shared his distress with him.

Not knowing what to do, William simply listened to everything while silently and repeatedly calling on Jesus to help. Josh sank into a deep silence for five or ten minutes ('resting in the Spirit', as William was later to realize). Then Josh suddenly shook himself, as if waking up. 'The shadows have gone!' he proclaimed. 'I'm free!' He was filled with joy and amazement. 'I feel wonderful—the dark, spectral figures, shadows and voices have gone!' Before, Josh had felt paralysed; his legs felt heavy, and he was unable to get up or even talk; but somehow the power of Jesus had broken through and released him from the darkness, which never returned.

'What do I do now?' Josh exclaimed, and there and then he gave his life to Jesus as the Holy Spirit touched him with his peace. Thrilled and excited, he started to read the New Testament in his new awareness of the Spirit. Ever since then he has experienced a special joy, knowing that the Holy Spirit is always present and that the risen Jesus is more powerful than any demonic intruder. The evidence of this power was a constant reminder to William to give praise to the Father in every circumstance. Josh was able to use his artistic talents to present his experiences vividly to other students who might have been tempted to doubt that dark spirits exist and can attack us.

Interestingly, Josh found that, when he went on to art college, the Spirit of Jesus could use him there to free others from the powers of darkness. But he had to learn not to go too fast, and never to work on his own without fellowship and prayer support. He also found that, unless the person he prays for really accepts Jesus, some of the darkness will return. He has become very wary of going ahead of where the Holy Spirit leads.

Josh is now comfortable with spiritual warfare—using the 'sword of the Spirit' (Ephesians 6) against any dark dimensions. He used to hate church or anything to do with it—but then he came back, realizing that all his tension, hatred and anger had gone. 'Amazing!' he exclaimed. 'There's something there for me now, after all those bad vibes.'

Josh was one of a group of students in the college who came to raise and discuss questions about science and faith with William after lunch. The other members of his group thought that what had happened to Josh was very strange! 'It's almost as if we're living in two worlds at once: the one we can see and the one we can't—like *Behind the Seen*!' they joked. Josh went on to get a degree and to become a lecturer in Media Studies at a university.* He now has great insight into the world of the occult and warns others against playing around with Heavy Death Metal and spiritualism, through which uncontrolled power comes from 'the other side', to use his words. Theologian Clark Pinnock reminds us that evil is overcome by 'the confidence we have as Spirit-filled believers in the power of God to deliver us from [it] here and now'.†

A Spiritual 'Big Bang' and a Singularity

Josh had always been fascinated by Einstein combining three space dimensions and one time dimension to propose *four* dimensions of *space-time*, and his general theory of relativity of curved space which had led

* He currently works as a freelance graphic designer and animator.

† Clark H. Pinnock, *Most Moral Mover* (Carlisle: Paternoster, 2001), p. 134.

to the Big Bang theory of the creation of the universe. This was the first major pillar of twentieth-century physics.

'What triggered the Big Bang?' Josh's group of friends asked William one lunchtime.

Before its explosive creation 13.7 billion years ago the entire universe was inconceivably hot, dense and crammed into a space smaller than a subatomic particle. Under these extreme conditions gravity and the other three fundamental forces behaved in entirely unknown ways and could indeed have combined into one 'superforce'. But without a theory of how gravity reacts with these, it is impossible to explain what started it all off.

William's students were staggered by the awe and sheer wonder of our universe, and by the concepts of black holes and dark matter. They explored the existence of black holes, the end point of stars much larger than ours after they collapsed. Black holes are something like a cosmic jail that imprisons both matter and light, making them completely invisible to observers (they are seen only by indirect means). Matter from surrounding stars is sucked into a black hole under pressure remarkably like that of the millions of tons at Niagara Falls, which pour into the chasm below with no way of stopping the flow.

'So what is around and inside the Big Bang and black holes?' William once asked the then Astronomer Royal and Professor of Physics at Durham. His reply was, 'You can't ask that question.' Physics has no answer. The laws of physics themselves break down at that point—called a singularity, a mystery in itself.

Some scientists believe that black holes may lead to other dimensions, a mind-blowing possibility but one that is beyond today's physics, as the equations turn to nonsense. At that time we were still awaiting the so-called 'theory of everything'.

With hindsight, it seems as if Josh's deliverance from dark spirits was like a *spiritual* Big Bang and an entry point to new dimensions of the Spirit. This is what cosmologists would call a singularity—but it was a spiritual singularity. This was the first of the Holy Spirit's actions of healing in which William was involved. It changed both William and Josh! It was to be the first of many healings William was involved with: all were different, and they often took place over a period of time, almost like continuous creation! Now, however, William felt the need to train in order to become qualified as a Christian counsellor and healer.

Interlude: The Principal's Tale

'When did you meet the Holy Spirit?' some students asked William.

After Josh's encounter with the Spirit, Josh and William remained on a high for some weeks. He and his friends were intrigued by the possibility that Jesus was the real singularity. They found the ten proposed embedding dimensions needed to describe Einstein's four-dimensional curved space to be an interesting analogy to being embedded in the Creator/Father's love for us.

However, what they really wanted to know was how William had met this Holy Spirit which Josh so vividly described. 'Had you gone on a retreat?' they asked him.

'No,' he replied (although he was to lead retreats to Holy Island in Northumberland a few years later, as we shall see). 'No, it happened on a flight over the Atlantic to give a lecture at Pittsburgh University.' He recounted what happened in his own words.

A journey into the unknown

First, I met Jesus. As a teenager I had drifted around a couple of churches, but I found church pretty boring and it didn't make any impact. Then I went to university, and one day I found myself listening to a rather dry talk given by a law professor—but it grabbed me! For the first time in my life I heard that Jesus had died for me, to give me a way to establish real contact with God. I discovered that, in sending Jesus, God offered everything of himself, not to remove evil, but to come into people's hearts to help them overcome it, if they would accept his forgiveness. The extent to which Jesus loved me struck me with great power—I had never heard this so plainly explained in the churches I had tried. That evening I gave my life to Jesus, willing to be used in his service. The following morning I experienced the joy of his presence as I read the Bible with new eyes and new understanding. I found its power as a creative word through which Jesus spoke to me. I still find this same power today, fifty years later!

Over the pond!

It was some years after this, while still a principal, that I was invited to give a paper to an international conference in America. I took with me a book about the healing of memories, which I wanted to read because I still did not feel at peace with my father about things that *had happened many years before, and I was upset that I had not been able to say goodbye to him properly when he died. I opened the book, praying, as I had done for some weeks, that I would be able to hear whatever the Spirit might want to say to me. What the Spirit did was remind me of how Jesus was with me at the time when my father was dying, and of Jesus' love for us both. Suddenly, I felt enabled to forgive my father and also to recognize where my own actions and attitudes towards him had been wrong. My pride in my own achievements and my academic degrees—even my pride in the paper I was going to present—crumbled. I realized that knowing Jesus and having his forgiveness were far more important than any of these things. It was as if an enormous burden had lifted. The Spirit gave me the gift of tears (a surprising gift—'men don't cry' in my culture, and boxers certainly don't cry—I used to be a boxer for Cambridge University!). I also had an almost physical, tangible awareness of Jesus' presence. I felt completely renewed and healed. From that moment I became conscious that I could invite the Holy Spirit to be present at any time—and in any place. Now I understood what Paul Y. Cho meant when he said that 'the anointing of the Spirit that comes through prayer ... belongs to a higher dimension than just natural wisdom and understanding.'*[*]

And it's not just in himself that William saw the Spirit work. This book tells of the Spirit working in the lives of many other people, and that was to be even more amazing.

[*] Paul Y. Cho, *Prayer: The Key to Revival* (Milton Keynes: Authentic, 1984), pp. 43–44.

Chapter 2

OUT OF THE CAGE: WILLIAM LAUGHED AND LAUGHED AND WEPT IN THE SPIRIT

About nine months after the incident with Josh, William invited Russ Parker, Director of the Acorn Christian Healing Trust, to speak to the whole college on Founder's Day in the nearby Anglican church. In his address Russ talked about the need for forgiveness. 'Forgiveness is healing,' Russ told his audience. Everyone has hurts in their life—often from childhood days, perhaps from father or mother, brother, sister or fellow pupils. Ross emphasized that hurts can be confronted and broken when, through the love of Jesus, we are able to forgive.

William was alongside Russ, facing the college students, and he could see in their eyes—both students and staff—that they agreed with his words. Russ talked further with a small group afterwards, but as the whole college came out of the church some shook his hand but many were afraid to meet his eye. Though they had accepted what he said they found it difficult to come to terms with what it meant for them.

Moved by Russ's talk William planned to attend an Acorn conference. However, as a busy principal, raising money for a sports, arts and media centre, it was only a while later that he was able to do so. Here he describes what happened there:

> *I was sitting with some friends at a table during the coffee break. Then Peter Ward, a medical doctor, started what seemed to be just a normal conversation with me. However, it was to be in the form of powerful prophetic words.*
>
> *'You must listen and accept God's praise,' said Peter. 'Well done in all you do—but now let others take over and do well. You will need peace, for healing comes through peace. Stop running and trying to do everything yourself—what you have been involved in will all come out well. You will do greater things, for I am with you—just wait and listen.'*
>
> *I laughed and laughed and wept in the Spirit.*
>
> *Others turned to look, but the experience was out of this world!*
>
> *'Does this make sense? I don't know what it all means,' said Peter.*
>
> *'Yes!' I said. 'Yes!'*
>
> *Peter also challenged me with the words, 'We all need to be healed for someone else's benefit.'*
>
> *Gordon Clarke, another visitor to the college, later confirmed the words of Dr Peter: 'You are to be released from the cage. Do not worry; the things you are involved with will come to fruition. Let others take over and have the glory.'*

Preparation for Retirement

And so it was to be. The Sunday following the conference, William released to the Holy Spirit both the book he was preparing at that time and the new college sports, arts and media building—willing to let them go unless Jesus wanted them to happen. Within eighteen months he had retired, aged sixty. (The building—the Pursglove Centre—was in fact completed and William was invited to return for its grand opening by the Duchess of Kent.)

It was the same Dr Peter Ward who, with his wife and Revd Anne Black and her husband, set up the King's House in Dunstan, Co. Durham, a healing centre for the area. William was present on one of their healing days with Peter and a group of visitors who were invited to share any problems or pain. One lady said she had been hurt and made very upset by the offhand nature of a surgeon. Her husband had died on the operating table, but the surgeon had simply come straight out of the room and had completely ignored her.

After she told them this there was silence; then Peter just said, 'Will you forgive me?'

'Why?' she asked.

'I am a doctor, and I stand in that doctor's shoes,' explained Peter.

And this lady wept and wept, but she was restored to peace.

The Centrality of Forgiveness

Around this time William also attended a conference run by Ellel Ministries, an international non-denominational Christian counselling and ministry centre that focuses on healing and deliverance. William was on the ministry team of helpers. Delegates were invited to stand if they wished to give their lives to Jesus as Lord, and William stood up alongside Charlie, the cook for the conference. After a while, when the speaker, Peter Horrobin, the director of Ellel Ministries, had finished talking, William asked Charles if he felt right.

'No,' he replied, 'at least, not yet. There are too many unhealed hurts from my difficult childhood.' In fact, he was helped with these during the rest of the conference week by Peter Horrobin and his team.

The Birth of CFM

Some weeks before William retired as principal he invited Gordon Clarke to the college to talk to his lunchtime group of students, to challenge them to a deeper dimension of healing. With a trim, athletic figure, Gordon was full of energy and real spiritual wisdom, having a keen mind coupled with many spiritual gifts and great love. He had been a nationally respected educational psychologist and a university athlete. His wife had died of MS some years before; wheelchair-bound, she had been nursed by Gordon for many years. Gordon had been an atheist until his wife told him how 'Jesus came to me in hospital', giving her assurance about the future. Two years later, after his wife's death and in despair, Gordon himself had an encounter with the Holy Spirit and he surrendered his life to Jesus. He gave up being Senior Psychologist at Durham County's Psychological Services to work part-time, and he founded a counselling and healing ministry called Christian Fellowship Ministry (CFM). This was financed by voluntary contributions and became a great support in intercessory prayer. (As we shall see, perhaps this prayer and fellowship was to be the strongest 'mortar' for the spiritual cathedral William was to discover.)

Gordon's vision to establish CFM had been affirmed through prophetic words given during a meeting with Selwyn Hughes, a Welsh Christian minister well known for his many devotional books and the founding of Crusade for World Revival (CWR). Gordon and his two colleagues were seeking advice on healing the broken-hearted and 'setting the captives free'. Selwyn wrote about this meeting:

> I was staying in a hotel in Blackpool, and when I awoke about 7am, I felt constrained to open my Bible at random (not my usual custom ...). Immediately my eye fell on Acts 10 v.19: 'Three men are looking for you. So get up and go downstairs. Do not hesitate to go with them. For I have sent them.' Intrigued, I went downstairs, where I discovered that three men had been asking for me. I walked over and introduced myself. That meeting with these men proved to be a major turning point in my life—and theirs.[*]

Inner Healing

When talking to William's students Gordon spoke of exploring the spiritual landscape. He described Jung's initiative of moving outside naturalistic values and explaining everything in physical terms, which could hinder or even destroy the spiritual development of his patients.

What the students were to remember later was Gordon's description of the mind being rather like a pear-shaped iceberg:

[*] Quoted in *Every Day with Jesus* (Farnham: CWR), 28 September 2010. Used with permission.

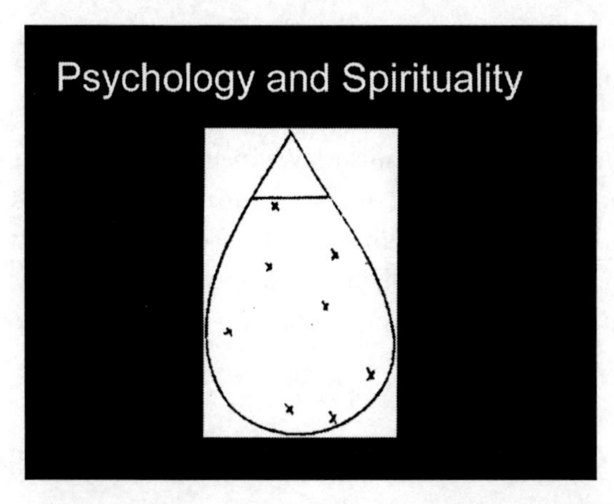

From the author's lecture at George Fox University, 16 April 2009, on his book-promoting tour of universities across the USA sponsored by Templeton Press

This diagram helped the students to visualize the other dimensions within themselves. The consciousness is only one-tenth of the whole. In the nine-tenths beneath the surface lurk the hidden memories of hurts (x) which need healing—which are 'buried alive', as Gordon put it. These must be brought to the light for the forgiveness and inner healing that are needed. The central issue is that forgiveness is essential—and that includes forgiveness of ourselves as well as of others who have hurt us on our life's journey. 'We can ask Jesus to walk back with us to the time when we were hurt,' Gordon explained. (Jesus was there anyway, as he is 'the same yesterday and today and forever', Hebrews ch. 13 v. 8.*) Taking our memories and bringing to light the things that have hurt us, and healing the binding effects of these, brings great peace. The alternative is to deliberately repress these memories and push them back down into the unconscious. Here they fester and cause us problems when something similar happens, triggering fresh pain and hurt. Instead, we can bring our painful memories under the healing and controlling power of Jesus. 'What is Jesus doing and saying? "Forgive them!"' Many find this hard. Yet seeing Jesus nailed on the cross, saying 'Father, forgive them, they know not what they do' shows us the way through.

Gordon pictured this process as someone visiting a house, deliberately descending the cellar stairs of anger in the company of Jesus and 'removing

* All Scripture quotations, unless stated otherwise, are from the Holy Bible, New International Version, NIV Copyright © 2011.

the pain as he integrates past and present. This will give peace as we again tread the floorboards above'.

He later spoke to the whole college of how we tend to deny our problems as a defence mechanism. We need to express our feelings if we are to know Jesus' love. If we don't do this we tend to displace our hurts onto others, without realizing it. However, if we allow ourselves to remember and forgive we will begin to move into the freedom of new dimensions and break the power of the past—perhaps even of our mother or our father—over us.

Stepping Stones to Accreditation as a Counsellor

Gordon's talk was William's introduction to depth psychology, but William needed further practical help towards becoming a qualified counsellor accredited by the Association of Christian Counsellors (ACC). Great encouragement came from an organization of doctors, consultants and nurse practitioners: Christians in Caring Professions (CiCP—now Forum for Christian Healthcare), and their conference 'Introduction to Christian Counselling' in 1998. William was greatly indebted to them for their training courses, conferences and video material. The wisdom gained by practising doctors and medical counsellors already engaged in physical healing seemed very secure, especially with the added spiritual dimension. As William explored inner healing, physical healing and deliverance the spectrum of ways to help people become whole began to widen.

One of his colleagues in a training group was Jim, a doctor. Jim later related that in his general practice he felt able to offer prayer at the end of a consultation. This was almost always readily accepted, even resulting in some patients transferring to him for this '*extra dimension of healing*'! Jim now runs marriage-guidance counselling courses with his wife.

At the CiCP annual conferences William was able to play the violin in the worship teams at the start of each day, which added another dimension in praise and joy. It was fascinating for him to find the leaders referring to 'spirits of anxiety' rather than just unpleasant feelings in the depths of our stomachs. William described the experience:

> *I was taken aback to be encouraged to break any spirit of anxiety in me (which there often was), to bind it and command it to leave—often to 'go to Jesus to be dealt with'. Feeling the Holy Spirit's touch, in the love of the Father, brought immediate peace. 'Does it*

work?' asked the scientist in me. 'Yes, it works!' was my wondering conclusion. And so I have had the knowledge and spiritual confidence to bring healing to others ever since—even myself, if I have allowed life to go too fast.

Soon William was able to test out further healing: 'generational healing'. When he asked if a spirit—perhaps of anxiety, infirmity or troublesome worry—had also been present in the father or mother, he discovered that this was often the case. Taking the 'sword of the Spirit' he was enabled to cut off any generational spirits and ask the Holy Spirit to come in at these points to bring healing, love, joy and peace to every dimension—body, mind and spirit.

These conferences confirmed William's reliance on the Holy Spirit in counselling, dealing with rejection, past hurts or false guilt through inner healing, and gave him a portfolio for the future. These approaches were particularly strengthening when engaged in spiritual warfare against dark spirits and entry points from the occult, using 'the armour of God' (Ephesians 6). Another useful weapon in binding spirits in the authority of Jesus were the words of Jesus himself in Matthew ch. 18 v. 18: 'Whatever you bind on earth will be bound in heaven, and whatever you loose on earth will be loosed in heaven.' The following verses are especially apt when working with other Christians: 'Again, truly I tell you that if two of you on earth agree about anything they ask for, it will be done for them by my Father in heaven. For where two or three gather in my name, there am I with them.'*

The Search for Unification

William's students were looking for a unifying principle of forces and particles in science as well as in inner healing, and they remembered from their earlier physics what were thought to be two completely unrelated forces which were seen to be really the same: magnetism and electricity. James Clark Maxwell's equation for uniting these two forces was one of the pillars of the modern age: electromagnetism as light or radio waves is accepted today without thought (this would be the first pillar of William's spiritual cathedral).

* 'This is the same power for the disciples as was given to Peter earlier.' Raymond E. Brown, Joseph A. Fitzmeyer and Roland E. Murphy, (eds.), *The New Jerome Biblical Commentary* (London: Geoffrey Chapman, 1995), p. 662.

'Let's Try Five'—'Does It Work? It Works!'

William had told his students about the lecture he gave to an international conference in the USA on his first visit there; it was about 'Kaluza and the Origin of Five Dimensions'. Theodor Kaluza was an obscure German academic who was trying to unify the two known forces at the time—gravity and electromagnetism. In 1919 he succeeded—using five dimensions.

William had visited Hanover to meet Kaluza's son, also a professor, and he related how he remembered the exact moment of his father's inspiration. He was eight years old at the time and was reading quietly in his father's study. Suddenly, Theodor Kaluza, Jr., said, his father stopped writing, 'was still for several seconds, whistled sharply and banged the table. "It works!" He stood up, motionless for a while—then hummed the aria of the last movement of Mozart's *Figaro*.'

Kaluza had found that the equations for gravity and electromagnetism were identical—*if five dimensions were used, and not the usual four.* How was he to publish this? He excitedly wrote to Einstein in April 1919 to enlist his support for this idea and received a postcard reply: 'The thought of achieving this unification through a five-dimensional world had never occurred to me and may be completely new. Your idea is extremely pleasing to me.' After backtracking, Einstein wrote again to Kaluza, two years later, to apologize and to accept Kaluza's idea of unification. Kaluza was overjoyed at Einstein's encouragement and published the paper. But the idea was so far ahead of its time—by fifty or sixty years—that it caused little stir. The modest and unassuming Kaluza died in 1954 without ever achieving the recognition he deserved. The challenge of this fifth dimension that can't be visualized needed more tools—and a new world picture that only emerged in the 1980s. The Kaluza model is widely used today, as scientists were to discover the need actually to use ten or eleven dimensions to unify the four forces of nature, building on Kaluza's original discovery that more than four dimensions were needed.

The arches to unify the forces connecting the great cathedral pillars of science were beginning to become visible: the spiritual pillars of unification through healing and wholeness were not far behind ...

Chapter 3

'HE'S OFF THE WALL!': A FURTHER CHALLENGE

Physical Healing: Another Pillar?

William had invited a local Baptist minister to talk to a group of interested students about physical healing, being quite agnostic about it at the time. A slim, athletic figure, Alan was a very wise and spiritual man. He was willing to share his own experiences of physical healing—such as of a nurse's arthritis and of a deputy head's back. His desire was to let the Holy Spirit in—'even to the churches,' he said with a wry smile, 'to invite God to let his healing power come on a person'. Alan wisely warned that there might be no apparent healing; however, the ***minimum*** result would be that the sufferer would experience a blessing. 'Sometimes,' he also cautioned the group, 'soaking prayer is needed over a period of time.' He also told them 'that *every* Christian ought to be prepared for healing'.

It was true!

This seemed quite challenging—even threatening—to the group! Alan went on to describe how, when the Holy Spirit came on him for healing, he felt a heat and a tingling in his hands, and that this was often transmitted to the person he was praying for.

'He's off the wall!' William thought to himself. He didn't really believe Alan, but went on listening anyway to the results Alan described. He talked of making Christians bold in speaking about the gospel and able to back up their testimony. 'God is around here in healing,' he said.

Later, as a scientist William explained, 'I was to find out for myself: it was true!' He said,

> *Sometimes I am asked when it first started for me. What a risk it seemed just to ask, 'Can I pray for your shoulder ...?'*
>
> *I believe it began when I started to attend healing conferences run by John Wimber from America. There was no set formula used by churches, but many of them came to experience a new level of expectation of God's working in power. John was a man gifted very powerfully with the presence of the Holy Spirit and he was happy to pass this on. At the time he was suffering from the throat cancer that would lead to his death some years later. He was in the habit of simply asking the Holy Spirit to come, and then waiting to see what would happen. Instead of telling us to bring folk forward for him to heal, John invited us to form small groups and to offer prayer for one member of our group. In my group of three or four, Ruth, who later became a fellow member of CFM and who was herself gifted in prophecy, told us that she had a very painful back. We asked the Holy Spirit to come and bring healing to Ruth's spine and back. Ruth felt the warmth from the Holy Spirit touching her as the pain diminished and she was able to stand and walk freely.*

Brighton Beach and T. S. Eliot: A Vision for the Future

At another conference run by Ellel Ministries in Brighton on deliverance from dark spirits and spirits of trauma and infirmity, William began to discover a deeper way of healing from physical injury. After he was encouraged to pray for the knee of a friend, Mike, the pain went and Mike was able to walk freely. Mike confirmed this result a year or two later, as well as the healing of chest pains.

It was while at this conference that William stood early one morning on the pebble beach, looking out to sea and remembering the vision of God's heart for China that Hudson Taylor received at this very place. William wept in his spirit for joy at what the Lord was doing in China—and was inadvertently 'baptized' up to the ankles; his feet were washed for Jesus!

While on the beach he watched the seagulls circling round and landing on the still water thirty yards out, away from the 'chaos' of the breaking waves which washed the pebbles and then retired for a while; there was a quiet point to retreat to even in a world of chaos. One gull was flying back and forth along the waves, lost in the morning mist. 'Stay calm where you are, with the hurts and the pain, until I tell you,' the Lord seemed to be saying to William. Pondering this, William returned to the hotel—it was still only 7 a.m. He entered the empty lounge and, to his complete amazement, heard a voice from the radio saying, 'Remain "at the still point of the turning world"', words from a poem by T. S. Eliot!* No one else was around and the words seemed so apt. 'What are you saying to me, Lord?' William asked, as he often did. The words on the radio were followed by music; it seemed that this had been just for him, confirming the vision he had received on the shore.

Only Prayer Mattered to Jimmy: From Secular Counselling to Prayer Ministry

William's experience of counselling grew from the honesty of Jimmy, a hospital anaesthetist. At Jimmy's request they had talked and shared for an hour about his inner hurts and compulsions, and they followed this with prayer. When William asked Jimmy which part of their time together had been most useful Jimmy replied, 'Only the last ten minutes when you were praying!' This changed William's emphasis in helping people: he was directed towards not just physical healing but also the dimensions of prayer ministry. Jimmy often came back to talk with William and in time he found a Christian fellowship where he felt at home.

'This Changed My Life!'

Richard, a young man from William's local fellowship, had been quite miserable and depressed for some time, and he also seemed very lonely. An elder and the youth pastor asked William if he would talk to him. A mild-mannered, caring person, Richard admitted that he had been having trouble with his boss, who had been unreasonable and frequently said hurtful things. He was unable to see a way out, so William asked him if he was willing to take Jesus back to the time of one of these painful situations. Richard said he was. William assured him that Jesus was bringing peace to the situation, but asked if Richard would forgive his boss. This was

* T. S. Eliot, 'Burnt Norton II', in *The Four Quartets* (London: Faber & Faber, 1986), p. 15.

hard for him to do. However, remembering Jesus dying for him on the cross and his forgiveness of those who were crucifying him, Richard was prepared to forgive all the hurts; William could see the depression seep away from him. It needed a further prayer session to bring deep healing and to restore Richard to a peace and love for others which never left him. Later he wrote to William to say, 'This changed my life!' He subsequently found a very fulfilling post within a Christian organization.

A Potential Author Meets the Holy Spirit

At around this time William was at Bridlington with a group of friends on a 'Holy Spirit Day' as part of an Alpha group which had started at their church. A Chinese-Malaysian preacher, Y-An, who was leading the conference, invited anyone who wished to come forward for healing. A rather gruff man—an ex-policeman as it turned out—came forward from the back row and shared some of his fears and worries. Y-An placed his hands near Graham's chest and invited the Holy Spirit to come. After a minute or so Graham found himself lying on his back on the floor. He remained there for about half an hour, 'resting in the Spirit', as it was called—resting in another dimension! After getting up Graham insisted on returning home immediately. His wife phoned the next morning before Graham rejoined us. 'What have you done with him?' she asked. 'He's a different person, without fears or phobias! He even went to bed and slept without the light on!'

This began a new life in the Spirit for Graham. William next met him when he was ordained as a curate at Whitby Abbey. The local witches he encountered (there are a number of covens around Whitby) apparently found the Holy Spirit within him to be more powerful than their own dark powers. Later, as a vicar in North Yorkshire, he discovered he had an ability to write about such things. Not finding a publisher for his novel about the battle between good and evil Graham sold his motorbike and self-published the book.* It was an enormous success and was later made into a film.

The Harbour Vision

It was as William was resting in the Spirit at this conference that he found he was unable to move from his chair. He explained:

---

* G. P. Taylor, *Shadowmancer* (London: Faber & Faber, 2003).

The Holy Spirit was in fact touching me and releasing further memories for healing.

'You're very quiet,' said the organizer, who wanted everyone to move their chairs and get into discussion groups. This didn't appeal to me! Gathering my coat, I went out for a walk—in a torrential downpour—and was taken by the Spirit away from the town to the harbour at Bridlington—in a gale-force wind! There I saw that the fishing boats and sailing yachts were crashing into each other, their masts clattering and their rigging jangling in the wind. However, they were at anchor and safe, while outside the harbour walls the storm was raging at sea.

'What are you saying to me, Lord?' I asked.

The Lord replied, 'Many people are like these boats and need healing outside the churches. Later on, the storm will be calmed and you will be able to open your sails and go out on the path I have chosen for you.'

Next day I took this word to my fellowship for confirmation of the vision. Albert, my long-time blacksmith and farmer friend, replied, 'Yes, but people need healing within the church also!'

When William visited the town again two weeks later, the storm in the harbour had long gone. The tide was coming in, floating the beached boats, and the fishing boats were sailing off for their next catch of fish.

Healing of Childhood Hurts

William found the Holy Spirit touching people for healing in all sorts of ways. While at the Ellel conference mentioned earlier, the sister-in-law of William's friend Mike was listening to the talk about how forgiveness can bring healing when suddenly she started laughing from the depths of her spirit, uncontrollably for several minutes. William asked Mike's wife what it all meant. 'It was all the deep hurts and pains from a very unhappy childhood suddenly being healed—most of them in this sudden explosion of emotion,' she replied. There had been a lack of childhood laughter in their very strict family life, and her sister had blocked this off in her memory. The compassion of Jesus now brought her releasing, restoring laughter. As Mike, his wife, sister-in-law and William held hands afterwards for prayer she frequently doubled up in laughter—a preparation for deliverance from the spirit of fear. Mike's wife had also been healed gradually over a year or two of forgiving—but her sister had, in one touch of the Holy Spirit, experienced a deep forgiving of all her hurts!

Jane's Tale

On another day at this same conference the delegates were invited to forgive those who had hurt them. The first suggestion for those they might need to forgive was young, thoughtless 'yobs', a particular nuisance for some older people. Jane, a lady in her sixties who walked with a stick, came across to the group William was with. The speaker had invited people to speak out their forgiveness, perhaps to others who didn't know them if they felt more comfortable with that. Jane admitted that, quite without personal reasons, she hated the youth of today. She asked for forgiveness as she confessed this to William's group. The group assured her that she was forgiven in the name of Jesus, and also asked the Holy Spirit to bring healing to her hip, which had been causing her some pain. As soon as she felt forgiven the pain began to disappear. Jane became a new person, full of joy and peace. She announced that, as a matter of faith, she would throw away her stick. She then walked with the group, somewhat unsteadily at first. Later, she was seen walking down some steps, still peaceful and rejoicing.

Jake's Tale

Another similar request for healing came the next day from a young man called Jake. Too embarrassed to tell those who knew him, he came to William saying that he thought he had an unclean spirit, a spirit of lust. William first asked Jake if he was willing to repent and ask Jesus for forgiveness, and then he asked the Holy Spirit to come in and drive out any spirit of darkness. Jake began to cough and cough—almost as if something deep within him was being expelled. Paper tissues nearby ensured that this could happen without any fuss, impervious to the strangers around him. It was wonderful for those who witnessed the release Jake experienced: to see his tense fingers relax and the joy of freedom come over him as he gave thanks in Jesus' name. 'Keep walking in the purity and wholeness of the Spirit,' William encouraged him as he returned to his friends.

Needing a Language to Talk About Dark Spirits: The Dark Dimension

At the same conference Bishop Graham Dow spoke about having a language in which to talk about healing in the power of the Holy Spirit, in the authority of Jesus, and also about dark spirits, dark powers—even demonic powers, as Jesus called them. Most people are no longer comfortable with deliverance from dark or evil spirits, and he encouraged those present to use biblical words where appropriate.*

However, discernment was needed in the use of such healing. Once William was back home in his fellowship Dave asked him for healing for his back, which had been painful for some time. Taking Jesus back to when the pain had started William asked Dave whom he needed to forgive.

'Well, no one—I did it through lifting too heavy a weight,' said Dave.

'Can you forgive yourself—or even God?' William asked him. This was easy for Dave, as William commanded any dark spirits of trauma or infirmity to lift up from him, and then asked the Holy Spirit to come and fill all the sinews, muscles and tissues with his healing power. As William held one hand a few inches above Dave's spine, Dave felt a flow of heat bring warmth and comfort (William felt this in his hand too). Dave was full of joy, giving praise to Jesus as the pain receded, and he was able to stand upright and walk around the room freely.

Dave came back to William a few months later. 'Can you heal my thumb?' was his next request. However, something made William pause: he said that it did not feel quite right.

'How did it start?' William asked, as usual.

'It was through playing my clarinet,' Dave replied.

William's discernment led him to ask Dave to check first with his clarinet teacher, because he was probably holding the instrument in the wrong way—and so it proved to be.

* See Graham Dow, *Explaining Deliverance* (Tonbridge: Sovereign World, 1991).

Science and Spirit

Making wider connections: mystery and paradox

The awe and wonder of seeing the Holy Spirit act in different situations seemed to William to resonate with the second great pillar of twentieth-century science, the quantum theory. It was Heisenberg's uncertainty principle that helped to challenge the three-dimensional Newtonian mechanics of the nineteenth century. A particle such as an electron is sometimes a particle and sometimes a wave. This wave/particle duality applies to everything and may well reflect the Trinity of Father, Son and Spirit. The quantum theory emerged in the 1920s: the quantum enigma surfaced as the theory was seen to involve the act of observation. Even conscious observation turns probability to certainty.* This is physics' 'skeleton in the closet'. There are other interpretations of quantum theory today, but each one encounters consciousness. Quantum mechanics is the basis of much of our technology and all our electronic equipment, yet it is quite mysterious. Quantum objects can simultaneously be in two locations and are best described by a probability wave in space. In 1927 Louis de Broglie suggested Pilot (or Guiding) Waves—he was ahead of his time, but William found this to be a wonderful analogy for the Holy Spirit. This is why we never know what state an object is in until we observe it. This is the paradox of 'observer-centred reality', seen with hindsight as 'Jesus-centred reality'! Friends of William's, such as Dave and Jimmy, asked him, 'Who was the observer of the Big Bang?' It's a valid question but one most physicists are reluctant to answer. Certainly the mystery of consciousness plays an essential role in the nature of reality. Material objects require consciousness in order to 'appear'.†

As William's Baptist supervisor, the late Durham Professor Euan Squires, used to say, we have been afraid of introducing any idea of God into physics. It is right to avoid using God to fill the gaps in our knowledge, but we don't have to eliminate the idea of a Creator God completely from our thinking. Two limericks summarize the problem quite neatly. The first is by Ronald Knox:

> There once was a young man who said 'God
> Must think it extremely odd,

* Bruce Rosenblum and Fred Kuttner, *Quantum Enigma: Physics Encounters Consciousness* (London: Duckworth, 2007), p. 6.

† Ray Tallis, 'Consciousness Not Yet Explained', in *New Scientist*, 205/2742 (2010).

If he finds that this tree
Continues to be
When there's no one about in the Quad.'

To which Bishop Berkeley, an Anglo-Irish philosopher who denied the existence of material substance, could have replied:

'Dear Sir, Your astonishment's odd,
I am always about in the Quad,
And that's why the tree
Will continue to be
Since observed by, Yours faithfully, GOD.'*

A new theology of quantum theory: entanglement and 'spooky action'

Einstein disliked mystery and uncertainty: 'God does not play dice with the universe,' he famously said. Yet in a 'thought experiment', two photons are allowed to fly apart to opposite ends of the universe. As soon as we measure the spin of one of the particles, said Neils Bohr, one of the originators of quantum theory, we know the spin of the other, no matter how far away it is. Bohr argued that this doesn't apply only to the microscopic realm, but also to objects large enough to be seen with the naked eye.

Einstein found quantum mechanics disturbing. He scoffed at what he called 'spooky action at a distance',† but theoretical proof (discovered in the 1960s by John Bell, a British physicist working at CERN, Geneva) was confirmed by French physicist Alain Aspect in the 1980s and others since then, overturning Einstein's view. Aspect's experiments could not be explained without action taking place at a distance, however 'spooky' this action might be. Often called 'entanglement', this universal connectedness is paralleled by a pillar of Celtic Christianity, the Celtic Knot, as we shall see. 'It's an unbreakable spiritual lifeline' to the presence of Jesus.‡ Nothing 'will be able to separate us from the love of God', as Paul states in Romans ch. 8 v. 39.

These mind-blowing paradoxes are endemic within quantum theory. As Richard Feynman, who won the Nobel Prize for Physics, commented, '[It's] screwy—a mathematical absurdity—like one hand clapping—Anyone who isn't taken aback, hasn't appreciated the mystery at the heart

* Cited in J. M. and M. J. Cohen, *Penguin Dictionary of Quotations* (Harmondsworth: Penguin, 1960), p. 225.

† John D. Barrow, *Theories of Everything* (Oxford: Clarendon, 1991), p. 195.

‡ Hebrews ch. 6 v. 19 in *The Message,* Copyright © 1993, 1994, 1995, 1996, 2000, 2001, 2002 by Eugene H. Peterson.

of the universe.'* The mystery of needing to invoke the act of conscious observation also points to the deeper mystery of consciousness itself—which in fact requires a spiritual dimension. As Professor Susan Greenfield confirmed, 'It is the ultimate puzzle to the neuroscientist.'† The enigma is still with us. Euan Squires often remarked during William's lectures and supervision classes at Durham in 1984, 'Every interpretation of quantum mechanics involves consciousness.'‡

We needed a new way out of these dilemmas—to unite the two disparate pillars of relativity and quantum theory—and this was to come in a world view of more than three dimensions ... The pillars of twentieth-century science have remarkable parallels in the mystery and wonder of inner healing and physical healing; perhaps faith is 'living with uncertainty' and needs a language of the many dimensions of the Spirit.

---

* Richard Feynman, published in *Six Not-So-Easy Pieces* (London: Allen Lane, 1999), p. 1.

† Susan Greenfield, *The Human Brain: A Guided Tour* (London: Phoenix, 2000), p.192.

‡ See also Euan Squires, 'Quantum Theory and the Relation between the Conscious Mind and the Physical World', Philosophical paper: 'Synthese 97', 1993, pp. 109–123.

Chapter 4

'I'M A HEATHEN ISRAELI; CAN YOU CURE MY ANGER?'

The Jew's Tale: It Was To Be a Long Journey

It was after William's retirement as principal. At a Christian Institute conference in Stockton-On-Tees Jim stood up at his table. He had been irritable with and defensive towards other participants. Somehow he appreciated that William empathized with some of his concerns. Jim shook William's hand warmly and asked him to cure his anger—a challenge to a duel, it seemed to William! Little did William realize, as he agreed, that this would result in a journey of questioning that would take an hour or so most weeks for almost a year.

A thickset, well-built man of striking presence, Jim in fact suffered from a very painful disease affecting his limbs. He had been taken to the Christian Institute by his wife, Chris, who had been praying for him. He phoned William the day after the conference and they agreed to meet at William's college. Meanwhile, William fasted and prayed for him; he knew that this was not going to be an easy time. Later Jim described their times together like this: 'William dared to challenge me, and our discussions were quite a battle.' However, the two became very good friends as they took walks together to the seaside or the park.

'Jim the Jew' had been a brilliant athlete at his secondary school and had represented his county in many activities. Jim was born of a Jewish mother in the mountains by the Sea of Galilee and on her side he came from a line of rabbis and philosophers. His father's background was quite different: he was a fitter in the steelworks, and his own father had been a spiritualist who led his own 'church'—a 'satanic gift', as Jim called it—he could never profit from it. Jim received no warmth or affection from his grandparents on that side of the family. Jim's brother had been envious of Jim's sporting ability and had gone to grammar school, later becoming a rich and famous barrister.

Rebellious and anti-authoritarian, Jim suffered frequent daily beatings from his father and resented his brother's academic success. His anger was directed towards everyone in authority. He and William had interesting discussions about this. Jim felt that God was punishing him now in many ways for not listening to him in the past. He had been having visions of transcendence, such as one time, when riding his bike to Hartlepool beach, the sunlight broke through the clouds to shine just on him. Later, he was transfixed by the image of a face on a rock until light set him free to move away. There were other occasions of transcendent wonder which, as the two talked, Jim recalled that he had avoided thinking about further. He wasn't ready to acknowledge that someone (maybe God?) was trying to speak to him. Owing to his Jewish inheritance he had not been allowed to read the New Testament. He could only manage to call God 'Big G', and at first only liked to talk about 'little j' (Jesus).

A Messianic Jew

Jim asked to be taken through Isaiah and there he found an answer: 'forgiveness'. It seemed so hard, yet it was at the root of his bitterness. He wanted to try to forgive his father, who got in the way of his knowing *the* Father. William challenged him to become a Messianic Jew and to study Isaiah further, and Jim readily agreed. William told him about the professor of Padua, who had refused to look through Galileo's telescope lest he have to change his world view.

After discovering that Paul had seen shadows and foretastes of Jesus in the Old Testament, they were able to discuss the power and authority of the Holy Spirit in the name of Jesus. William mentioned Joel and the prophecy 'I will repay you for the years the locusts have eaten',* a promise to restore all Jim's past. Although William told Jim of his having renounced his science degrees and also his pride in being a headmaster/principal before he could know the touch of the Holy Spirit, Jim tried to

* Joel ch. 2 v. 25.

counter that it was *because* of William's education that he listened to him! (William also told him of missionary and professor Sir Norman Anderson's words many years ago, that Jesus died for *him*—and that Anderson had kept his faith* despite the death of his son at sea, aged 21.) Jim had heard Jesus speak to him through William, but he was still running away.

Fishing boats offshore

A picture Jim saw, of fishing boats beached at Saltburn, signified him waiting for the tide to turn. Jim knew he was on the right path and he felt the peace of the Holy Spirit in the pub where they had coffee. He knew he needed to repent, to come to terms with the stumbling block of the cross that bridged the gap to Jesus. Only the Holy Spirit could show him Jesus as Saviour and Lord. The burden for Jim was now lifting; it was as if he was beside the Wailing Wall of repentance and prayer in Jerusalem, coming to know Jesus and weeping for his lost pride. Through reading Acts 9, they saw Paul knocked off his horse and similarly humbled before meeting Jesus.

'Can we run this course together?' Jim asked. 'Only if we walk the Jesus way,' William replied. Jim asked if he could join William's fellowship—'When the time is ripe,' William promised, but he agreed to bring him in prayer at the next meeting. Jim still needed to accept authority, but he had been much moved and challenged, beginning to feel like the prodigal son of Luke 15. The forgiving of his family continued as he cut off generational ties, especially from his spiritualist grandfather, and stood in their shoes for forgiveness. There were also signs of healing in Jim's knee after William prayed and broke any spirits of infirmity in him, but the rheumatoid arthritis was still painful.

The fellowship weekend, and 'Jim takes the plunge!'

At lunch a week or two later Jim apologized for all his anger. He was a new person and healing was in progress; he said, 'I'm very close to giving my life to Jesus.'

The culmination came at a weekend led by CFM, under Gordon Clarke's direction, at Wyedale in North Yorkshire. Jim had been quite ill and someone gave up a single room downstairs for him. A dreadful night followed a talk on 'The Road to Calvary' by David Woodhouse, who led the worship the next day, centred on the cross. Jim was anointed with oil

* J. N. D. Anderson, *Christianity: the Witness of History* (London: Tyndale Press, 1972), p. 104.

and later recounted how the following day was a 'resurrection day' for him, when he praised and sang to the Lord.

This was to be the centre of his testimony when Jim was baptized at Stockton Baptist Church some months later. He was full of the final moment of release from any satanic powers, such as spiritualism, by the powerful name of Jesus. Jim had found there a Christian fellowship of committed brothers and sisters. He told the congregation how he remembered when the Spirit first spoke to him and he wouldn't listen. 'Jewish independence, even arrogance,' said Jim, now tempered by his Christian humour. He was to buy a New Century Bible—'the most significant book I have ever laid my hands on'.

This had been a hard road for him, yet he had enjoyed the mutual challenges as his friendship with William developed. One prophetic insight came when the Spirit nudged William to tell Jim ('out of the blue') about Naaman. In order to be healed, Naaman had to immerse himself seven times in the river Jordan—'a foreign river' for Naaman, who at first refused (2 Kings 5). Jim laughed and laughed as the Spirit spoke to him and foretold his own water baptism.

'Jim Thwaites takes the plunge!' Jim wrote on the poster he designed later to advertise his baptism. His wonderful, caring and long-suffering wife, Chris, had had her prayers answered at last!

William's task had been to introduce Jim, despite his Jewish background, to Jesus and to ensure he joined a fellowship. William now had to take a sabbatical from walking with him at the CFM weekend as the challenge of a teaching post beckoned him out of retirement.

Postscript

Jim subsequently applied for a postgraduate course. The last stage of his pilgrimage came when he phoned William to say he had got his MA in Creative Writing from Edinburgh University. This achievement of his goal came two years before his death at home.

The Irish Catholic Folk Singer's Tale

Like Jim, Rick's was also a long journey of healing from past hurts. A retired psychologist, Rick was a tortured soul, and there was much forgiveness needed for his early schooldays. He was regularly beaten on the knuckles by a very stern teacher—and this was made worse because the teacher was actually his father, headmaster of a Catholic Brothers school.

Anger and resentment had entered Rick's spirit and led to his need for ongoing prayer for the healing of his memories. 'Forgiving till the pain goes' marked his long association with William.

Ill-suited to becoming a priest, but undergoing training at a Catholic seminary, Rick fought bravely against temptations. He was badly advised by another psychologist to let his anger come out; this produced much hurt in his family. After many 'jump starts' in the Holy Spirit Rick knew he was forgiven and always phoned William to give thanks. As he came out from the guilt-ridden demons of his religious background he began to know the healing power of Jesus and the reality of spiritual warfare against dark forces.

He greatly enjoyed holding evenings of Irish folk music with his wife and was very proud of his family. Rick remained a wonderful and loyal friend; sadly, William was unable to be at his funeral in Ireland, which became a celebration of his life in true Irish style.

A Parallel Pillar: Inside the Atom

Some of Jim's challenges with William had strayed into science. He had learned about atoms being like small billiard balls, the smallest individual particles of matter. Once scientists had moved beyond this they needed a better model for the interior of the atom. The one which was developed in 1913 and came to be known as the Rutherford–Bohr model was based on the planetary system. Around the nucleus of neutrons and protons, electrons orbited like planets around the sun—later, to Jim's excitement, to be described as waves in multidimensional space. The Rutherford–Bohr model was itself superseded, but it made the point that actually everything consists of empty space. Jim was intrigued to hear of the paradoxes of the quantum theory and especially of Eddington's 'two tables'. Eddington was fond of pointing out that there are two tables, one on which he leaned his elbow, and the other of theoretical physics—mainly empty space. 'If you think of the nucleus as the size of your thumb,' William conjectured, 'the rest of the atom, the part with the tiny orbiting electrons, would be the size of a football field—or indeed Durham Cathedral.' Most of the atom consists of empty space. 'And if a table is largely empty space, so is your elbow and even your brain,' laughed Jim. He was keen to know where it all ended: was it like the Russian doll William's son had brought back from North-West China, with a smaller doll inside, and a smaller doll inside *that* doll, and so on?

Quarks make sense of the jigsaw

It was in 1964 that Murray Gell-Mann developed a beautifully simple model that would explain many of the difficulties. This was based on the idea that the protons and neutrons are not solid but are themselves built up of smaller particles, which he called 'quarks'. These apparently fundamental building bricks were named, in his whimsical way, after the line 'Three quarks for Muster Mark' in *Finnegans Wake* by James Joyce. At first these were just mathematical non-visualizable, untestable ideas, ridiculed by professors of physics. But within ten years (to Gell-Mann's own surprise) the model became more and more useful, was increasingly accepted and has now become standard.

Quarks, strings and many dimensions

Yet no one has ever seen a quark, and there is still no direct proof that quarks exist! Jim was very thoughtful at this link between science and faith. The six different quarks have quaintly named properties of up, down, strangeness, charm, top (or truth) and bottom (or beauty); perhaps these are best thought of as extra dimensions or degrees of freedom, rather than physical properties. Jim was himself moved by the same curiosity and invention that inspired scientists such as Gell-Mann.

Jim and William were reaching the limits of their discussion in this area, but they needed to take a peak even inside the quark doll. Physicists had been focusing on the apparent conflict between the pillars of relativity and quantum theory to search for an overall theory of everything to connect and unify these two areas. It was to be found in strings, not particles—in fact, superstrings in ten dimensions. This reminded Jim and William of the Holy Spirit: though unseen, he is the key to wholeness and true reality.

That was enough for Jim, but they left the ideas simmering as they looked at the parallels between science and spirituality. Jim was always ready to try to find a language in which to talk about the paradoxes and mysteries of Jewish and Christian religions. He remained euphoric after catching the idea of extra dimensions—maybe they are 'dimensions of the Spirit', he prophetically exclaimed.

Chapter 5

MARTIN, THE MOTORCYCLIST WITH SEVEN VOICES

'Can You Help Me Find Peace?'

It was some time later that, now as an accredited counsellor with the Association of Christian Counsellors (ACC), William met Martin the motorcyclist. After a phone call to arrange an appointment they met one morning in the newly opened Pursglove Centre. Martin's wife had refused to live with him after he had crashed his fist through the locked bathroom door. He had agreed to go to the doctor and was referred to one psychiatrist after another. By those willing to accept him he was prescribed psychotic drugs. Martin had refused to take them as 'they blew my mind'. He felt he was being biochemically changed with unwelcome side effects and he wanted to be in control of his own life. It was at this stage that a friend recommended he come and talk to a Christian counsellor.

Martin's appearance was very intimidating. Over six feet tall and dressed in dark leathers, he entered carrying his black helmet, which he would not allow anyone else to touch. He sat down at William's invitation, refusing to remove his black leather jacket or to release his grip on the helmet. However, it soon became apparent that his leathers were both a protection from the world and a shield behind which he could hide himself. At some personal risk he became willing to talk about the voices which were troubling him: anger, jealousy …

So began a journey which was to last some months. Was Martin willing to go the Jesus way, going back into the past? If not, William would use Cognitive Behavioural Therapy (CBT) and other secular methods. However, his experience was that only by working and praying in the power of the Holy Spirit would Martin be able to obtain release into the peace which he craved. Secular counselling would help for a while, but it was a bit like putting sticking plaster onto a deep wound which really needed spiritual cleansing and healing.*

'When Did It All Start?'

They began with Martin's traumatic birth. He was one of twins, the other of whom was stillborn; his mother had been so weak that she had had to be rushed into hospital. After the twins were born Martin's grandmother poked them both and found that one moved ... Martin was still alive, though he needed much care. There were to be many other disturbing experiences in his life, and he and William had to look at each one, taking Jesus back to see what he was doing and saying about them. Once, locked in his bedroom, Martin destroyed all his toys; on other occasions he was pushed downstairs and thrown into a door. Martin found he didn't like remembering.

'Which voices shall we tackle first?' William asked, looking for the entry point. Martin chose to remember being locked as a small boy inside a tin trunk by four or five bigger boys. His fright turned to rage, and this anger had long been repressed, covered by the voice or spirit of anger. This spirit often erupted later, usually triggered by an apparently harmless event. Martin was willing to see Jesus there (as he had been all the time, although unrecognized and not called for). He saw Jesus unlocking the chest and standing between him and the older boys, with his arm around Martin. 'Are you willing to forgive them?' Jesus asked Martin. This was a hard question. Martin and William talked of Jesus dying for us, confronting evil on the cross and crying, 'Father, forgive them, for they do not know what they are doing' (Luke ch. 23 v. 34). When Martin found himself able to forgive he and William could then drive off the spirit of anger within him with the words 'Get out in the name of Jesus!', commanding the spirit, by name, to go to Jesus, never to return, and then asking the Holy Spirit to come in and heal these wounded places in his memories.

---

* See Thomas R. Insel, 'Relapse Rates High', in 'Faulty Circuits', *Scientific American*, 98/2 (2010), p. 35.

The Journey Towards Healing

Martin knew immediately that this spirit had gone. He felt released but knew that he still needed to be freed from other voices. So continued his journey. Martin would choose which voice to tackle next, and he sensed at which point each one had come in to shield his conscious mind from a particular traumatic experience. He described 'the depressed one', which had been in charge of all the other voices, 'like a black maggot eating inside'. 'Mr Anger' was triggered by frustration or annoyance, but there were also 'Mr Flirty', 'Mr Sarcasm', 'Mr Cynical' and two others. The most destructive voice was Jealousy, which had to be dealt with after Anger.

As David Woodhouse, psychiatrist, counsellor and vicar, so graphically described such an experience at a Christian conference, Martin's life was like a sandwich cake, divided into seven slices, with one slice extracted at a time. The current 'filling' needed to be healed and removed so that the Holy Spirit could 'fill' the slice with his love, peace and joy. The slice could then be put back into the cake and a further wedge removed for healing. Sometimes Martin knew that two spirits/voices had come out at once, perhaps jealousy and lust, leaving him in a state of peacefulness and released from the dark dimensions.

Public Testimony and the 'Ship's Engineer'

Although Martin appeared to have a tough protective shell he was in fact a gentle soul inside. After some weeks he allowed William to touch his helmet, and he returned to his biker comrades in *white* leathers. Martin found he was now able to read the Bible and to look at verses in the New Testament, always asking the Holy Spirit to reveal himself through the Word. From time to time he was led to share with his friends how he had been healed, and he memorized scripture verses to replace what the voices had been saying to him. He still constantly needed to forgive those who had hurt him in the past, and also to forgive himself for things that had happened. His tendency to judge himself too harshly was healed as he took Jesus to all the dark places, one by one.

There came a time when Martin was ready to speak out to others in our fellowship group of the deep healing he had received. 'It was,' he testified, 'like being a waterlogged ship, sunk beneath the surface of the harbour.' He had needed 'a ship's engineer,' he went on, to restore and heal the painful memories that were shut off and lurking beneath the seven voices he kept hearing. William confided, 'I was quite proud to be named as the ship's engineer, remembering my wishful answer to my headmaster when asked, aged ten, what I was going to be. "An engineer, sir," I replied, never dreaming that this would become true in a most unusual way.' After speaking to the fellowship group Martin wrote to William, 'Hello, my great mate. Doing testimony was odd but was quite rewarding.'

Martin and William had become good friends and still keep in contact. Martin tried a variety of jobs, such as motorbike instructor. He often returned to William's house to chat about his progress—sometimes about having slipped back and needing to renew his walk with the Lord. At one time he became landlord of a local pub and found he was able to welcome church house groups there for meetings and discussions. More recently, he was working at a centre for young people with epilepsy and needed much prayer, as Christians in social work are often viewed with suspicion. 'You are a great role model,' he wrote in an email to William a year or two later, 'and I have my unshakeable faith because of you.' He prayed for the Lord's will and blessing in William's own life: 'be safe in the knowledge that you have made a difference on this fleshy earth.'

'Did anyone in *your* family have a problem with anger?' Martin asked William during one visit. 'Why yes,' William admitted, and he told him the story of his great-grandfather, John.

Victorian Psychiatrists: Great-Grandfather John's Tale

John was a much-loved, kind and respected family man. He had been manager of an ironstone mine in Lincolnshire and came of very independent stock. His father, William, had taken the Queen's shilling and had fought in the battle of Waterloo in 1815, after which he was bought out of the army by his parents—only to take the Queen's shilling and enlist again! (As the Duke of Wellington had famously ordered his troops when they defeated Napoleon, 'Stand firm!')

John's wife died when he was about 60. In those Victorian times John felt unable to share his feelings and to talk about his grief, about what his wife had meant to him. Quite uncharacteristically he became violent, even to one of his daughters. This resulted in his being taken to a mental asylum in nearby Lincoln. Asylums in those days adopted a variety of medical treatments but they were no more successful than the bloodletting and purgatives that were gradually abandoned. New treatments at that time included using chemicals with strong side effects, as well as ECT (electroconvulsive therapy). Doctors could not agree on these treatments, which were 'more likely to do harm than good'.*

Just talking introduces another dimension of healing. It was very fortunate that John was in the care of a psychiatrist who was ahead of his time in treatments. No ECT or padded cell for him; instead, he just began to play cards—cribbage—and to talk with John. After a year or so John was released and able to come home.

When William's own father and sister went to visit John, they were warned not to mention what had happened; it was considered shameful for the whole family. However, John was so happy at having been able to release his pent-up grief that he took the initiative to say what had happened and how being able to talk about it had restored his sanity.

Men today often still have an inbuilt masculine reserve. William's former neighbour, Fred, found it too painful to talk about his wartime experiences as a naval engineer on merchant navy ships on the 'Arctic Run', and like so many seamen suffered even more from these repressed memories. However, it has often been a particular privilege for William to be involved in leading other men into healing of their memories.

Martin rejected drugs for psychosis, and Chartered Clinical Psychologist Richard Bentall has commented, 'Let's listen more and drug people less.'† He states that conventional psychiatry has been 'profoundly unscientific and also unsuccessful' at helping the most profoundly distressed and vulnerable people in our society. Cognitive Behavioural Therapy (CBT) is better than other methods, he writes, but at best is only modestly effective when used to treat the symptoms of psychosis. He emphasizes that warm collaboration and relationships are the key to success in psychiatric care. Martin himself said at the start of his meetings with William, 'I feel better already just through talking.' William needed to respect Martin's own efforts to understand the events in his life experiences that had led to his breakdown, and to treat his beliefs seriously. Years later, Martin was again asking for prayer for a new job and encouraging William to keep on: 'You fix people; of course you're going to be used in healing.' William's willingness to befriend Martin met his need of unconditional acceptance—he took a risk in disclosing the unacceptable, the voices in his head—and offered both hope and opportunities for recovery.

---

* Sarah Rutherford, *The Victorian Asylum* (Oxford: Osprey, 2008).

† Richard P. Bentall, *Doctoring the Mind: Why Psychiatric Treatments Fail* (London: Allen Lane, 2009), p. 264.

Needing A Spiritual Dimension for Progress

David Woodhouse is still in touch with William and recommends a number of books for working with Dissociative Identity Disorder. However, with all of these different approaches, he wrote to William in October 2009, the spiritual dimension will need to be addressed, 'and that is for the Christian therapist to weave into the process'. Psychotherapist Phil Mollon refers in Valerie Sinason's book to 'the dark dimensions of multiple personality'* and describes sudden shifts from one personality state to another, often sudden and sharply defined (as with Martin). This book also has a chapter on Post-Traumatic Stress Disorder, in which schizophrenia is seen to be often brought on by trauma. CFM also helped in Martin's dark days of sliding back, which were emotionally exhausting in the voyage of unexpected discovery.

For Martin, as with Jim earlier, anger was not the root problem but was a result of far deeper experiences of hurt and pain. Francis MacNutt, a world-renowned preacher, healer and author who spoke at a conference at Harrogate, emphasized that it is these roots that Jesus wants to heal. Often this anger arises because people have been deprived of love and treated unjustly. 'Most evil spirits oppress people because of wounding that has happened in the past.'†

A Further Pillar of Science Today: Strings in Ten Dimensions, Superstrings and More

Both Martin and William are lateral thinkers, looking for connections between apparently unrelated ideas. They had often talked about joining the completely different pillars of science: relativity and quantum theory. Gravity doesn't fit in with the other forces we know about. So far, our best explanation comes from string theory, the leading candidate for a 'theory of everything'. Martin was intrigued to find that string theory requires that the universe has more than the three spatial dimensions that we experience—possibly as many as ten or eleven. The theory suggests that the unseen dimensions are hidden from view because they are rolled up small, compacted, or 'end-on' to our familiar dimensions, making it hard to detect their presence. They may also explain why gravity is so weak compared with the other forces; according to string theorists, gravity, unlike the other forces, leaks in and out of these extra dimensions. We only get to experience a fraction of the true strength of gravity.

* Valerie Sinason, *Attachment, Trauma and Multiplicity: Working with Dissociative Identity Disorders* (London: Routledge, 2009 repr.), pp. 110–117.

† Francis MacNutt, *Healing* (London: Hodder and Stoughton, 1989), pp. 228–229.

Martin was euphoric when he and William reached this point in their discussion. 'It's just like the Holy Spirit! We need to ask the Holy Spirit into our hearts each day'—the reason being, William chortled, was that 'We leak!' Martin enjoyed the new analogy of the Holy Spirit being present in many dimensions. 'But of course he is!' he exclaimed. 'Heaven is in God's dimensions all around us!'

He and William were beginning to accept the strangeness of having more dimensions than the three spatial ones our common sense insists on. Perhaps this experience resembles the difficulty with which people accepted Copernicus' theory and Galileo's proof that the earth went round the sun when they had assumed the opposite. It was against common sense—like the extra dimensions of string theory.

Superstring Theory: Many Dimensions in Science and Spirit

String theory provides a unified theory of the universe, given that the fundamental ingredients of nature are not particles but tiny one-dimensional filaments called strings. String theory unites the great pillars of early twentieth-century physical science, quantum mechanics and general relativity, which are otherwise incompatible. Michael Green, appointed in 2009 to the Lucasian Chair of Mathematics vacated by Stephen Hawking, discovered with John Schwarz that the problems he was working on suddenly fell into place. This was named the 'First Revolution' in superstring theory, 1984. He proposed that subatomic forces and particles were tiny strings, vibrating in space and time, differing merely in the ways they vibrated, like the different notes on William's violin. Michael Green once said that he could think of the universe as a symphony or song, made up of the notes produced by strings vibrating in different ways.

William once visited this professor and asked where these extra dimensions were. Pointing both hands towards him and moving them in clockwise circles Michael replied, 'They are all around you and within you.' As Martin commented, 'That's just where the Holy Spirit is—in extra dimensions of the Spirit all around us!'

String-theory research is carried out in every university, said Michael—it is a thriving subject, for unifying forces and particles, and 'there is nothing else'.

There remained the challenge of holding these extra dimensions in our minds ...

CHAPTER 6

THE GOATKEEPER'S TALE, AND TALES FROM OTHER EX-COLLEAGUES IN A PREVIOUS SCHOOL

'I Was Adopted, You Know!'

William often stopped to chat in a nursing home on his way to visit his family further north. Pat was a long-term patient who had been a former colleague of his at a school where he had been headteacher; she had been a most caring head of the Rural Studies department. A real country-lover, she had obtained approval to keep a small herd of goats there. Her work had been enormously valued but she seemed unaware of this and needed to be reassured. Often pupils from deprived homes were delighted to be able to help her with the goats; they enjoyed being able to touch and stroke them without any fear of the reprisals they were used to in their tough backgrounds. For them it was like entering another dimension. Children who have been abused will often show love to animals they consider will not hurt them or let them down. Pat had been very fond of these 'Goat Club' pupils and was really thrilled when William was able to bring her their latest club magazine.

Pat was happy to talk about Jesus and affirmed the truth of a favourite analogy of William's, that often rocks are thrown into our river of life. These come through no fault of our own, but are hurts from others which stop the flow of our spiritual journey.

It was after one visit, as William was just leaving, that she burst out with the words, 'I was adopted, you know!' He hadn't known—and her story revealed deep hurts in her past. The key to her healing was 'Emma'. When aged about eight or nine Pat and her fellow schoolgirls were playing rounders one day. Emma was the keeper behind Pat and she suddenly shouted at Pat in a spiteful way, 'You were adopted, you know!' She told Pat that her mother and father were not her 'real' parents. Pat remembered cycling home feeling terribly lonely, 'as if the sky had fallen in'. Her adoptive parents had not told her. Emma and the other girls hated her, she realized, and they wanted to hurt her.

Pat was a gentle spirit but during her life she had suffered frequent rejection—by her biological parents, by her Polish husband, whom she had married when he was in a POW camp and who left her after the war ended, and by her adoptive parents, who no longer wished to support her—and this had contributed to her arthritis. There was much forgiveness needed. She and William had talked earlier of the Goat Club magazine and Pat was able to try the daily Bible readings that were recommended in it. She had not known the love of the Father heart of God.

Deep forgiveness is the central key

Above all it was Emma, the root of all her hurts, whom she most needed to forgive. First she had to meet Jesus and know the depth of his forgiveness of her. 'He died for *you*,' William gently told her.

'But didn't he know me?' she wondered.

'The Father knew you even when you were in the womb,' William was able to assure her. She was amazed at Jesus dying for *her*, and at the wonder of his love. She had been reading in her New Testament that Jesus forgives. 'It sounds too good to be true!' she exclaimed. And yet she agreed that it was true, and she asked if CFM could pray for an increase in her faith. She discovered that we ourselves can be forgiven as we forgive others, and we can be set free.

A nurse, Mary, entered the room at this point.

'Did you watch *Born To Be Wild** with Martin Clunes last night?' she asked. He had been taking the 26-year old Nina, a tame elephant, back

* Shown on BBC1, 1 September 2002.

into the wild. He needed to tempt her first; she hesitated at the gateway to the wild country as if unsure, after all her life in captivity, that she *wanted* to be free! 'This is for you, Pat!' William said. Pat had hated Emma, but now she was willing to take Jesus back and express that hatred. For a long time she had been unwilling to forgive Emma, but now she had heard Jesus say, 'You must forgive her' because of the cross. She was then ready to forgive Emma (and even bless her!) and any others whom the Holy Spirit brought to her mind: father, mother, husband and daughters. This also included many at her high school who had constantly derided her for being adopted, even the 'godly headmistress' who hadn't wanted her in her school because it brought the stigma of adoption upon it. William was able to pray with her each time as the gift of tears came.

She now wished William to call her Patricia as he tried, unsuccessfully, to trace her twin sister, who was adopted separately at the same time. William did find out that Pat was actually christened Faith (and her sister, Hope) by the National Children's Orphanage at Harrogate.

Pictures from the Holy Spirit

Christian friends at CFM often prayed for Pat and gave William two pictures for her which Ken had received. She knew that the way of forgiveness is hard, and Ken's first picture was of a crown of thorns—but with fruit within the thorns. Another time William told her of Ken's picture of a rough road leading to a well. There was a bucket inside the well, and when it was pulled up, a notice became visible: 'Forgive'; through forgiveness the road on the other side of the well became smooth.

These 'words from the Lord' had a very peaceful effect on Patricia, as she pondered the love of her heavenly Father and her—indeed *their* adoption (William's too)—by the Father. She had become much brighter and her speech, which had been difficult, improved: she spoke more clearly and cheerfully, as she was living now in hope and love. 'It all fits together,' she was later to say. She heard Jesus saying 'I am with you' as she felt the presence of the Holy Spirit in her room.

She died some months later after feeling much better in her spirit. William was sad but rejoiced that she had come to know Jesus. Mary, the nurse, phoned William to say that Patricia had died peacefully in bed, so William lifted her spirit to the Lord, to her real Father in heaven.

Asked to say a few words at her funeral, William told of her accepting Jesus through the healing of memories, and of Ken's prophetic pictures. 'She was a real pilgrim,' said William, and he asked the Holy Spirit, in the dimensions all around us, to speak into the hearts of all who were present at the funeral of Patricia Faith.

The Senior Tutor's Story

Perhaps it was through hearing these words at Pat's funeral, and knowing that William was now an accredited counsellor, that Brian phoned him on behalf of his wife, Pam, a senior tutor who had been a colleague of Pat's and William's and had been visiting Pat at different times. Pam was seriously ill; her legs were becoming paralysed and she was to be admitted to the local hospital. William prayed using the centurion's prayer from a distance,* asking that Pam be healed. 'We will be holding her in our prayers,' he assured Brian, not knowing that he was to be personally involved in the future.

Pam then asked if William would visit her in hospital. Her illness, which was possibly due to a viral infection, had already lasted for two months and had resulted in asthma. Pam had been a superb teacher, first as head of PE and then in a pastoral role as year head. Her optimism and cheerfulness had always inspired all who knew her. William felt honoured that she had asked him to visit her.

First, William needed to know what had triggered this debilitating illness, remembering that Jesus had asked the father of a demon-possessed boy, 'How long has he been like this?' (Mark ch. 9 v. 21). It transpired that Pam thought she was to blame for not looking after her mother-in-law more (she had died three months previously). This was so unlikely, as Pam was always a most helpful and kind person. William and Pam agreed that in fact Pam needed to bring forgiveness to herself. 'If there is any possible guilt,' William reminded her, 'Jesus' death on the cross brought forgiveness, and false guilt needs to be rejected.' Giving her a prayer for evening time William also prayed with Pam against any spirits of trauma or infirmity. William commanded any such spirits to lift up off her in the name of Jesus, and he also prayed against any side effects of the drugs she was being prescribed. He asked for the healing presence of the Holy Spirit to enter Pam, to heal every fibre and nerve in her body, and also to heal any viral infection.

On a later visit William needed to reject any spiritual darkness that was causing her nightmares and send it to Jesus, never to return. William and

* Matthew ch. 8 v. 8: 'But just say the word, and my servant will be healed'—prayer from a distance.

Pam again invited the Holy Spirit in, so that they could pray for others and perhaps sing choruses she was familiar with from her Anglican background. At her request he also gave her Ephesians chapter 6, on the protective armour of God, and John 15, on the power of love. Later, they were to break any spirit of anxiety or undue worry, as from the ACC conferences William had received the power of the Holy Spirit to break any spirit of anxiety. Some anxiety can be motivating, but debilitating anxiety can be bound and rejected in Jesus' name. William shared with Pam his own experience, which was to cut off any generational spirits of anxiety in Jesus' name, taking the sword of the Spirit to break their effects. He left Pam, feeling exhausted but joyful in being able to be a spiritual catalyst for her. She looked brighter-eyed and cheerful as they said goodbye.

Meeting the presence of the Holy Spirit

In William's subsequent visits to the hospital ward he was to bind and reject any spirit of fear for an unexpected operation on her foot and any residual spirit of anxiety. They talked about the book of Acts—really the Acts of the Holy Spirit in bringing peace and love and joy, which William had been studying in preparation for a future teaching request. There was continued prayer as Pam now knew the Holy Spirit was present, especially in the long night hours, and she was able to share her pain with Jesus, who she knew had died for her. This helped her to relax, release her worries and receive the healing power of the light of the Holy Spirit within herself.

At a lunch given for ex-colleagues from the comprehensive school those present said how much Pam appreciated William's visits, calling him her 'Father Confessor', a reference to Pam's High Anglican upbringing. When William explained about the healing of memories with Pam's friend Jean, Jean exclaimed, 'Scary! You never know what Jesus may want of you, or where it might lead!'

A fortnight later Pam had made excellent progress and William continued to pray not only that she would walk again, but also that she might claim the victory of Jesus to walk in the Spirit each day! Pam did indeed gradually improve enough to walk again and to be released from hospital. When he spoke to her on the phone William found that she was very much better. He felt able to share how much the Father loved her, and he asked for the Holy Spirit to bring continued healing and peace. Pam admitted that she had felt driven and guilt-ridden all her life; she had always undervalued herself and her remarkable gifts. Now, however, she saw the need to forgive herself and remove any false guilt. 'Don't try too hard,' William said laughing, 'just welcome the Holy Spirit. No more guilt! He'll come just to bring healing and peace.' Three months later Pam was still making excellent progress, and could even walk upstairs.

Later, William was invited to Pam's 'welcome' into her local Methodist church. They have remained close friends. Like Patricia earlier, Pam found out how much she was valued and loved by the Lord: 'You cannot earn his love or lose it: you are very precious in his sight,' William said to her, paraphrasing Isaiah ch. 43 v. 4. Pam and her husband are currently acting as hosts for Ethiopians on the Global Exchange* programme. She recently emailed William to say, 'I am extremely well, involved in church life, and of course keeping fit.'

The Agnostic Guitarist's Tale

Another of William's ex-colleagues who asked for prayer for healing was Keith. He was a dynamic, much loved and highly respected teacher and a long-time friend of William's. Athletic and vibrant, Keith was a folk dancer in his spare time and he often enlivened his lessons with his guitar and creative folk songs. Much sought-after for in-service training of other teachers, Keith was always full of life and humour.

The day he asked for prayer he was feeling quite depressed; he and William had been talking about his family and the school. Keith admitted that his neck and shoulders had been extremely painful for some time. William recounted, 'I felt that in the dimensions of the Holy Spirit I was being challenged to pray for Keith (the first time ever!).' So he tentatively suggested, 'You need some healing prayer for your neck!'

'Can I pray for your neck?'

An avowed agnostic, Keith nevertheless replied, 'I'll have anything!' When asked if he would like William to pray for him he immediately replied, 'Yes, please!' As usual, William warned that he might not be healed immediately but that it could take some time, and Keith readily accepted this.

They moved to a private room and William asked Keith how it had started. Was there anyone else to blame? 'Only myself,' Keith replied, 'for pushing myself too hard.' Taking Jesus back to the time when it began William said that Jesus was asking Keith to forgive himself for going too fast. In Jesus' name William broke any spirit of infirmity or trauma. Then he simply asked the Holy Spirit to bring his healing and peace to all the

* Visit www.globalexchange.org/programs/

bones, sinews and muscles in Keith's neck and shoulders ... William began to feel the warmth come from his hand, which he held close to Keith's neck, vertebrae and shoulder muscles. Keith could feel the heat, often a sign of the Holy Spirit penetrating the body. After an additional prayer based on Ezekiel's River of Life allegory in Ezekiel 47, William told Keith of the need to walk in forgiveness, in peace and in holiness—and he agreed!

'It's better!' Keith exclaimed after a minute or so.

'Take care!' William reminded him. 'It's often a process needing soaking prayer.' Keith knew this and walked free as he went through the door, though William remembered to hold the heavy outer doors open for him!

William writes, 'We were often to meet again from time to time and it was always a joy to remember, sharing friendship and recent experiences.'

Ed Witten and Duality in Eleven Dimensions

Keith was extremely interested in how contemporary science had been looking for a theory of everything. This was to be the central pillar of the progression from the incompatible relativity and quantum theories. He had always been intrigued by the notion that reality was greater than three dimensions of space. In 1996 Ed Witten of Princeton University used the principle of duality to prove that the five different theories of superstrings in ten dimensions were actually the same underneath, looked at in the right way. He named the overriding theory of everything as M-theory in eleven dimensions, where M stands for Membrane, Mystery, Magic or the Mother of all theories.*

Keith's lateral thinking immediately saw how this 'duality' could be applied in theology to the Trinity: Father, Son and Holy Spirit were all one, as he indeed had been discovering in practical healing!

Ducking out—or a new language?

'How on earth can we hold this in our minds?' mused Keith. We needed a language to think and talk about such transcendent ideas. He knew William's story of a famous batsman, captain of England at the time who, William recalled, 'after hitting my first ball for six into the next field, hit the next ball so hard back at me that I hardly saw it, never mind got my hands up to catch it.' (The batsman was David Shepherd, later Bishop, and William bowled him out next over, as Keith liked to remind him.) Was this

* Ed Witten, 'The Pied Piper of Superstrings', in *Scientific American,* Nov. 1991, p. 18.

like not being able to hold these *ideas* in our minds? The progression of William's spiritual cathedral pillars, now brought together in astounding fashion—in eleven dimensions of M-theory—was certainly hard to grasp.

The extra dimensions of space were to be widely accepted in M-theory, 'the greatest thrill of my life', Ed Witten had exclaimed: 'it works!'*

Evidence for superstrings!

There were those who asked for concrete proof before accepting M-theory. It is widely believed that the Large Hadron Collider (LHC), a particle accelerator now operating at CERN, underground near Geneva, will provide the evidence for superstrings. Yet as Martin Rees, Professor of Astronomy and Astronomer Royal, has affirmed, it will also vindicate the vision of scientists from Einstein to Witten. 'To me', he wrote, 'criticisms of string theory as an intellectual enterprise seem to be in poor taste. It is presumptuous to second-guess the judgement of people of acknowledged brilliance who choose to devote their research career to it.'† The Atlas Experiment at CERN has further confirmed the presence of the Higgs particle. This then confirms the Standard Model of particle physics to describe the universe (proposed by Peter Higgs and others in 1964). Another theory predicts that the Higgs interacts with corresponding 'Supersymmetric particles'.

Peter Higgs himself says, 'I am a fan of supersymmetry, largely because it seems to be the only route by which Einstein's gravity can be brought into the scheme'‡ – and if supersymmetry, then superstrings and the extra dimensions needed for M-Theory.

'That's good enough for me,' Keith said, 'but to hold eleven dimensions in our mind, we need to drop back to two'—we need to imagine ourselves in just two dimensions. Keith immediately thought of the prisoners in Plato's Cave analogy, where the shadows cast on the wall of the cave represent our own three dimensions. We ourselves are then only three-dimensional shadows of a deeper reality in many dimensions—certainly in deeper spiritual dimensions.

* Ibid.
† Martin Rees, 'Mathematics: The Only True Universal Language', in *New Scientist*, 3695, 14 Feb. 2009, pp. 36–39.
‡ 'Why it's nice to be right sometimes', interview in *New Scientist*, 21 July 2012, p. 28.

CHAPTER 7

'I'VE TRIED EVERYTHING FOR EIGHT YEARS,' SAID COLIN

The Musical Director's Tale

I've tried everything for eight years,' said Colin. 'The doctors and hospitals don't seem to be able to cure my back problem.'

When looking for a publisher for his first book William came to a conference held by the Association of Christian Writers (ACW) to meet fellow Christian authors. After a morning session Colin, obviously in discomfort, approached William to tell him about his back pain, wondering if he could help. He was a musician and a director of a musical. 'How did he know to seek me out?' William wondered (although in fact he had led that morning's worship). Not knowing what would happen William agreed and suggested that they had lunch first.

After William agreed to pray for his back they looked for a private place to go to. Colin was willing to take Jesus back to the time of the trigger. It had happened when he and a friend had lifted a heavy trunk down some stairs.

'Whom do you need to forgive?' asked Jesus. 'Is it a particular friend, yourself—or even God?'

Colin acknowledged that it was mainly his fault but he was willing to forgive both the friend and himself. For eight years he had thought about how stupid he had been. William was about to bind and cast off any spirits of trauma and infirmity when Colin mentioned that his father also had back trouble—and that he was a Freemason. William was aware that this false religion can be an entry point to dark dimensions through idolatry and terrible oaths taken in the rituals of initiation. It often has the effect of making someone unable to receive the Holy Spirit. William therefore took the 'sword of the Spirit' to cut off any generational ties and to forgive and ask for forgiveness in Jesus' name. Colin was willing to stand in the shoes of his father (also admitting some occult activities himself), renounce Freemasonry and ask for forgiveness. William and Colin finally cut off all ungodly ties to his dead father, from him to his mother, and also to his sister (who was now also into the occult). This still left any bonds of love untouched.

Holy Spirit energy

Now William was able to command any spirits of trauma or infirmity to leave Colin and ask the Holy Spirit to bring healing up from his feet and all up his back. (As usual, William was not touching him but holding his hand an inch or two away, knowing that the Holy Spirit was using the spiritual heat energy which came into his hands when he asked for healing.) Finally, Colin requested healing for his left ear and to be able to hear the Holy Spirit in a balanced way. He was much moved and at great peace.

'Why can't I do this myself every day?' he asked.

'You can—but be ready to heal others too,' William told him.

'I must go out for a walk,' said Colin, 'to test whether it really has worked ...'

He walked out seven feet tall. 'Take care,' William urged him, 'your back may need soaking prayer for a few days—and spiritual protection from the armour of God!'

The next day Colin also mentioned his knee.

'Just walk in the Spirit,' William said, 'and feel the Holy Spirit bringing healing and freedom for the knee, in Jesus' name.' Three days later, at the end of the conference, Colin came to give thanks not only for his back but also because his knee was healed and he could hear the Spirit guiding him. This was a one-off! Neither he nor William were ever the same again!

Some of the gifts of the Spirit include pictures, visions and dreams, which William had begun to see as wonderful windows in the spiritual cathedral, letting clear new light of revelation through from another dimension. These pictures often come when they are not expected. It was at the writers' conference, after Colin's healing, that delegates were invited to ask the Lord for a picture to share with others. William shared his harbour vision: that the storm in the harbour had now died down, the tide was coming in and the beached fishing boats were afloat, ready to sail out. Then he was given a picture of a swimming pool. A woman was at the shallow end, scared to move further down, but Jesus was taking her by the hand and leading her to the deep end. He then led her, from her first anxious steps, up to the platform on the high diving board. Rachel was amazed: this picture fitted her thoughts so well! How wonderful that William had been led to join her very group of four! In fact, Doris, on William's left, was also astounded; it confirmed a very vivid dream she herself had had, of standing on the lowest step of the diving platform and being afraid to climb higher.

William was to find that two members of his Alpha group accepted Jesus as Saviour and Lord because of his account of Colin and the swimming-pool pictures he had received from other dimensions. The ACW conference had been both a springboard to getting his book published and a source of prayer and fellowship—like the mortar used for building the cathedral.

Breaking Words Said Against Us

His counselling accreditation course helped William to explore another dimension of the Spirit: the power of Jesus in breaking words said against us. His first experience of this came after an operation for a detached retina, which was carried out in a local hospital by the wonderful surgeon who had already operated on a previous retina tear. A Spanish assistant consultant examined his eye the morning following the operation. She said nothing to William but went into a corner with the nurse: 'It hasn't worked,' William heard her mutter. Although it turned out later that the operation *had* been a success, these words produced despondency and depression in William. It was some weeks later, in an Ellel worship and prayer group, that Janet spotted that he was not his usual self. After he explained what had happened with his operation, thinking that this might be the reason for his change in mood, Janet was quite sure that these words had affected his spirit and she asked if she could pray for him.

'I break any power of darkness in the words said, and any spirit of infirmity in William's eye, in the name of Jesus,' Janet prayed. She asked for the protection of the Holy Spirit while the Spanish doctor had contact with his eye. The depression immediately lifted and William knew that a dark spirit had left him at this point.

Breaking cutting words

This made such an impact on his spirit that later William felt able to help out an old friend, Graham, who had retired from being a vicar some three or four years before. He had been suffering from nightmares, which disturbed the family's sleep. William travelled up to Caithness to visit him, and the two shared about a number of healings of memories from their past. Then Graham told William of the cruel, cutting words said by someone at his retirement. These words had eaten into his spirit and produced angry flashbacks whenever he tried to sleep. Graham agreed to take Jesus back to the time of his retirement and to feel how much the Father loved him and valued his work as a minister. Jesus also said that Graham would have to forgive the person who spoke the hurtful words. Graham was able to do this, and William and he broke the power of these words in the name of Jesus, asking the Holy Spirit to cleanse his memories and to bring peace and joy back into Graham's life. He became so much more his old self and, though he was still on some medication, his wife said later how much more peaceful he was at night. This proved a great blessing to all the family and to his relationship with William. They were able to claim the Lord's words in Joel ch. 2 v. 25: 'I will repay you for the years the locusts have eaten.'

'Only 2 o'clock in the afternoon': William shares his own story

It was the following year that I received a deep healing myself when Eric Delve spoke at a local church. As with earlier conferences soon after I retired as a principal, I was tempted not to go. Eric spoke so movingly and powerfully, saying 'God loves you'—that he died for me—that I went forward for anointing with oil (the only time I have done this). This was a dedication for service and for God's blessing. 'He will do a new thing,' Eric prophesied as he touched my forehead with oil. I stayed kneeling at the front rail, keeping my hands out to receive the Holy Spirit, when Eric came back over and anointed my hands also, for praying and for healing others. I was also commissioned in his service to write! The Spirit came upon me*

* Eric Delve, *To Boldly Go* (Tonbridge: Sovereign World, 1994).

in warmth and slight shaking. As I knelt further down to resist this I found I was keeling over to one side, just allowing the Holy Spirit to be present, which gave me a deep peace for some time. (I thought I could have risen at any time but I chose to allow myself to rest in the Spirit, as if I had lain down in a stream which washed over me.) Sometime later I staggered somewhat drunkenly back to my seat, allowing the Spirit to be present in every part of me. It reminded me of Peter's words in Acts ch. 2 v. 15 when the Holy Spirit came on them at Pentecost: 'These people are not drunk, as you suppose. It's only nine in the morning!'

I was in another dimension of the Spirit. Perhaps it was like the experience Paul had when he said that he was in the third heaven (2 Corinthians ch. 12 v. 2)—perhaps I was in the seventh, remembering M-theory with an extra seven dimensions! The vicar of the church asked me for a prophetic word. I could only say that the Lord works through wounded people—'wounded healers'. I drove back home slowly and peacefully and in awe of nature, as I was in love with Jesus again. Feeling called to visit the recently opened Pursglove Centre I took the blessings of the Holy Spirit with me—it was a praise day.

It is this phrase 'wounded healers' which so aptly describes the group of people who have been my inspiration in counselling and healing. I had earlier been welcomed and accepted into Christian Fellowship Ministry (CFM). This fellowship provided the strongest 'mortar' for the building of what was to become my journey into the dimensions of the Spirit. Very privileged to be invited, I was able to work with an accredited supervisor, which was essential for all the counselling that was to come, and to be part of a ministry team at different conferences. We still meet from time to time for 'away days' and intercessory prayer. Now, however, my time as a regular member of CFM was drawing to a close as new challenges were opening for me.

An open door to the Isle of Skye

However, during this time other healings were happening, including during a CFM mission to the Isle of Skye which I undertook with Jean, Thelma, and Gordon, our director. We had been invited by the Christian community in the south of the island to talk and lead discussions. We stayed in an old fisherman's cottage and found the islanders very welcoming and open to our speaking through parables of nature and the countryside. Imagine yourself, we suggested, as a flowing river coming towards a very steep waterfall. You must decide whether or not to risk going over the edge and down into an unknown future, refreshed for the next part of the journey. I had pondered this myself while standing at the top of High Force Waterfall in North Yorkshire. There the water is also able to run off into side channels instead of going over the top. However, in the dry season these side channels become murky and fester without any throughput of water. Were the people on Skye willing to be healed in the resurrection power of the Holy Spirit in risking giving their lives to Jesus? After High Force, I encouraged them, came the beautiful, peaceful course of the river to Low Force, with its lesser waterfalls.

Later, I shared my own experiences of making a parable film of the river Tees. From a mysterious hidden spring at the summit of Cross Fell the initial stream (our spiritual life) often goes underground. It emerges later to encounter waterfalls and the stored energy of reservoirs (prayer). A colleague and I hired a small plane to look at the river from above, filming the incised 'middle-aged meanders'. We joined a pilot boat for the final stage of the journey to the sea.

Windows from the Spirit: pictures brought a stillness as they felt the Spirit's words were for them

The islanders were fascinated also by the picture of the rough road to the well and the word 'forgive' brought up on a bucket. I said that this was originally given to a friend, Patricia (see Ch. 6), but 'if it feels right, it is for you also!' Alf came to me the next day saying that this was changing his life. He had 'taken' the card with 'forgive' written on it and put it into a crack in the wall surrounding his life. He was able to bring forgiveness to another person and the wall then collapsed: he was now free to go through.

Always the emphasis was on the presence of the Holy Spirit dimension. Many were Celtic folk who knew the presence of the Lord. It was Jack the pigeon fancier who told us how, when a homing pigeon was returning, he would hold out his open hand with grain and the bird would just touch his hand lightly. 'You can't grasp at it,' said Jack, 'you must just wait for the touch as it homes in.' Jack explained that the Celtic word for 'pigeon' was actually 'dove', and this seemed a good analogy for 'the dove of the Spirit'.

The real healing is wholeness, independent of any physical healing

It was the experience in CFM fellowship that although physical healing often happened, the real healing was wholeness of spirit, a healing from hurts and unforgiveness. This healing is much more than curing an illness: indeed, the final healing may come in dying, in the active hope of the resurrection. 'We needed to acknowledge the mysteries that there are in God's kingdom,' wrote Eric Gaudian, who had been holding on through twelve years of chronic pain. From a life submitted to Jesus he speaks of a theology of pain and suffering.*

Other friends such as Jean, Alan and one or two others in Jean's ecumenical healing group bring a conviction of healing others while living in the grace of God through their own suffering. Their attitude, like Eric Gaudian's, is one of powerful acceptance of their own situation, and they continue to have confidence in the love and goodness of God. Their faith, as Jean says, is 'joyful and persistent', irrespective of the mysteries of life and death; they are truly 'wounded healers', mortar for the spiritual cathedral, like other prayer support groups.

The Engineer and Cello Player's Tale

One of William's closest friends, Godfrey, a civil engineer by profession, was to end his life in great peace, free of pain in a hospice, surrounded by family and friends. 'As the glory of God is manifest in Jesus' suffering and pain, so in giving ourselves up to others, doctors and nurses, and waiting with Jesus, we experience a "new dimension of glory",' as W. H. Vanstone wonderfully describes 'the waiting'.†

A stocky, good-humoured man of great faith and love, Godfrey was the cellist in William's string quartet, which is one of the best examples of the many dimensions of music, especially the music of Haydn, Beethoven, Schubert or Mozart. It was a privilege for William to pray with him and

* Eric Gaudian, *Storm Force* (Milton Keynes: Authentic, 2009), p. 152.

† W. H. Vanstone, *The Stature of Waiting* (London: Darton, Longman & Todd, 2004), pp. 98–99, 'Dimensions of Glory'.

share some of his final days. They often felt the wonder of entering the dimensions of the Spirit while playing together, keeping in time, checking their A strings together, perhaps using a tuning fork with correct A note (their analogy for Jesus) and also practising individually. The different strings really resonate with each other: 'All the dimensions are covered in a quartet,' said Godfrey. 'I feel we have entered another plane of existence.' Some think the second violin or viola part is best: it is at the centre of the multidimensions of sound—superstrings or not!

Another branch of current science they talked about was chaos theory.

The Theology of Chaos Theory: Jesus the Real Fractal

Within the apparent chaos of all that had been happening in everyday life, another pillar of contemporary science seemed to be showing William a deep hidden order. This was the science of the multidimensions of chaos theory. As Stephen Hawking remarked, 'The whole history of science has been the realisation that events do not happen in an arbitrary manner, but reflect a certain underlying order.'*

Fractal images translate mathematical descriptions of the hidden order within chaos into a visual image. The apparently patternless behaviour becomes simple to interpret, because what looked chaotic is highly structured.

In *The Fractal Geometry of Nature*† Benoît Mandelbrot revealed the forms within nature. These were not just the visible forms but the beautiful hidden shapes 'embedded in the fabric of the universe'. Powerful computers are used to represent the multidimensional phase space by a fractal map. The complexity of many dimensions would have remained veiled in abstract mathematics had its visible nature not been revealed.

The cross can be seen as embedded in the history of creation. In the vision of parallel pillars Jesus can be described as a fractal image of God: the hidden pattern of the many dimensions of the love of the Creator/ Father revealed in the three dimensions of Jesus. We can't see the deeper aspects of reality until we have a way to perceive them. It is significant that in the story of Jesus stilling, even rebuking, the storm, it is described in the *New Jerome Biblical Commentary* as 'Jesus Conquers Chaos'.‡ In the Old Testament storms were often symbols of chaos.

* Stephen Hawking, *A Brief History of Time* (London: Bantam Press, 1988), p. 122.

† Benoît B. Mandelbrot, *The Fractal Geometry of Nature* (London: W. H. Freeman, 1982); also see Nina Hall, (ed.), *The New Scientist Guide to Chaos* (London: Penguin, 1991), p. 125.

‡ Robert J. Karris on Luke ch. 8 v. 24 in Raymond E. Brown, Joseph A. Fitzmeyer and Roland E. Murphy, (eds.), *New Jerome Biblical Commentary* (London: Geoffrey Chapman, 1990), p. 698.

Stepping Stones to a Deeper Vision

It has been William's own experience that although at the time he can only see an apparently random series of events, yet when he looks back he sees a pattern emerging. He was soon to see this reinforced in the journey he was travelling—one in which he never knew what the next step would be or what other pilgrims he would meet. More doors were soon to open into other dimensions of a spiritual reality.

Commenting on Luke ch. 4 vv. 35, 39 Karris writes, '... "rebukes" is a technical term from the vocabulary of exorcism of unclean spirits', p. 691.

Chapter 8

OTHER DIMENSIONS OF REALITY: 'CALLED BACK TO TEACH'

The phone rang: 'Happy New Year!' was the greeting from an old friend of William's, the college principal of Cuthbert's. 'Could you teach some theology for us?' she asked; her head of department was ill again. As a trained scientist and chemistry teacher this came as quite a shock to William. Pondering a moment he replied, 'Only if it involves a *science and faith* module'; he thought this would be unlikely! After a while a second phone call came: 'Yes! Can you start this Thursday?' This would fit in with the book he was working on. Not long afterwards the college also asked him to teach New Testament theology, and then Old Testament (these were subjects William had studied many moons before). Later, philosophy and ethics also emerged: these were exciting challenges and much preparation was needed for a scientist like William.

The different groups of students he taught were amazed at real postmodern science and to discover that there are many parallels between science and faith. Certainly there is no conflict, despite the view of the man in the street. The science these students had learned at school was little preparation for the mind-blowing excitement of today's science.

Science And Faith: The Story So Far

William and this group of students at Cuthbert's felt awe and wonder at the size and shape of our own galaxy, seen end-on as the Milky Way in the night sky, and the expanding universe of millions of galaxies. Like earlier groups of students William had spent time with, they saw how Einstein's relativity theory, the great pillar of the early twentieth century, laid the basis of the Big Bang theory of origins. This took the students to the presence of black holes in the universe, and even to the massive black holes at the centre of most galaxies, including our own. The strange singularities of the Big Bang and black holes were only one of the mysteries they encountered. Were they in fact entry points to other dimensions, as some physicists, such as John Barrow, Professor of Mathematical Science at Cambridge University, have suggested?* Some students saw embedding dimensions as a parable of God's love; others found Jesus to be the real singularity, the real entry point to the dimensions of the Spirit.

Mystery and paradox were seen further in the uncertainty principle: that particles can be in more than one place at a time. Quantum mechanics, though an exact description of the world of the very small, is almost too strange to believe. It led to the description of particles as waves of probability and so to observer-centred reality. Only when an observation happens does the probability wave collapse to reality. 'If an observer (or measurement) is essential for a description of reality,' the students said, 'surely Jesus is the real observer': Jesus-centred reality indeed.

They further encountered 'non-separability' of particles once originally joined but parted by a great distance—'entanglement', or, as Einstein described it, 'spooky action at a distance'. They noted that this was like the prayer of the centurion in Matthew ch. 8 v. 8 and saw William's 'China watch', worn on his right wrist and set to China time, as a visual metaphor for this (William's son and family had been in China for many years—and he was soon to visit them†).

They explored past protons to their constituent quarks. Originally postulated by Murray Gell-Mann in 1964 with the words 'Let's try quarks'‡ they were at first rejected by down-to-earth physicists, yet within

* John D. Barrow, *The World within the World* (Oxford: OUP, 1988), pp. 313–316; e.g., 'one can think of the entire Universe as the interior of a black hole', p. 314.

† See 'Time Out (1)'.

‡ Murray Gell-Mann, speech given when accepting the Nobel Prize for Physics in 1969 for his classification of elementary particles known as 'quarks'.

ten years they were accepted because 'it worked', although no one has ever seen a quark.

No one has ever seen strings either, the next layer down the current model to describe particles; the current theory of superstrings works only in more than three dimensions. As we have seen, it was Theodor Kaluza who in 1919 first said, 'Let's try five dimensions!' to unify the two known forces at the time. To his great joy and excitement the equations of gravity and electromagnetism become identical in five dimensions. He was ahead of his time. It was not until 1984 that Michael Green and others put forward 'superstrings' as the basis of a new theory of everything. 'At what cost?' William's students wondered. 'We need ten or even eleven dimensions for it to work, building on Kaluza's original idea.'

The groups of students were all taken aback by M-theory. 'Do we really live in an eleven-dimensional world?' The other seven, beyond our three space plus time, are, as Michael Green described them to William, curled up at each point in space. 'How do we hold these in our minds?' was the students' next demand. They needed a language to even talk about it. So with their imaginative help William explored the story of Flatland.*

Flatland Revisited: 'Bridging the Gap'

Entering Flatland for yourself

There are only two-dimensional beings in Flatland, a two-dimensional plane. They cannot see one another from above, only as a line from within Flatland, perhaps with an angle to another line for a triangle or square. Above them, but unseen from below, is the three-dimensional Creator/sphere, hoping to communicate with the Flatlanders. Each group of William's students described how this might happen: lifting one up, perhaps the square, and showing him the wonder and splendour of Flatland from above, then returning him so that he can tell the others. 'Do they listen to him?' Then perhaps lifting up a king or emperor with some effect, but one that doesn't last long. 'What does the sphere think? Does he destroy them all, perhaps with a flood?' Most students said that the sphere had invested a lot of time and care into the Flatlanders …

---

* Based on Eric Middleton, *The New Flatlanders* (Godalming: Highland Books, 2002). Revised and republished by the Templeton Press, 2009, and again in a Dutch edition, 2010. Visit www.thenewflatlanders.com/. *The New Flatlanders,* in turn was based on the nineteenth-century novel *Flatland: A Romance in Many Dimensions* (1884, UK Penguin Books, 1986) by A Square (alias E.A. Abbott).

Eventually someone would suggest that the sphere comes himself and takes the risk of entering Flatland. Of course, only the two-dimensional cross-section of the sphere would be visible as a circle. After his wise words, mending some broken sides, the court hierarchy would become wary of this apparent upstart with his group of supporters.

'What happens next?' William would ask the students.

'They will plot to get rid of him—kill him!' is the usual reply.

'But can you kill a sphere? Or have they just wounded a slice?' What happens next? Does the sphere abandon his followers, or does he return to show that they cannot kill him? Perhaps to enter as a Flatlander—a gardener, fisherman or traveller—perhaps just to enter a closed upper room, or, as most suggest, lift them all up, at least those who are willing!

At this point, although it is a completely secular story, someone would say about the circle, 'That's Jesus!' William would often mention Philip, the disciple who had heard Jesus give his followers the Lord's Prayer and who asked, 'Lord, show us the Father.' Jesus replied, 'Anyone who has seen me has seen the Father' (John ch. 14 vv. 8–9).

We are just three-dimensional shadows?

After some thought Dan erupted, '*We* must therefore be three-dimensional Flatlanders, shadows in a universe of higher dimensions!' (Some scientists refer to this as a hologram.) This was the group's best attempt to cope with the mind-boggling extra dimensions which seem to be needed in physics today. They were struck by a verse they found in the Gideon Bible that had recently been presented to the college: 'These are a shadow of the things that were to come: the reality, however, is found in Christ.'*

'Flatland really made me think!' many students responded.

'A new world view,' William challenged them; 'can you cope?'

These extra dimensions have a startling and creative use in describing the incarnation, cross and resurrection, where the dimensions of the universe were expanded. Instead of Jesus being confined to one place, Palestine, the Spirit of Jesus is free for the entire world to welcome him if they seek him. He is a gift; he is not to be earned in any religious sense. As William mentioned to the students, the only criteria for them to meet the Spirit were to renounce pride in any achievements and also to forgive, to repent of all unforgiveness, and to be able to welcome the Holy Spirit every day.

* Colossians ch. 2 v. 17.

Only Anth and Richard were unwilling to enter the story, though it was not told by William but elicited from the group. Perhaps they could see where it was leading and didn't want to be challenged? (William adds, 'It was with a "Flatlander" group in a local prison that Anth and I first met, and we have remained friends ever since.')

The Mystery of Dark Matter—and the Anthropic Principle

To complete the science and faith module William and his students talked of the mystery of dark matter and dark energy, which no one has seen directly yet which constitutes 90 or 95 per cent of the observable universe and is postulated to explain the rotation of galaxies. The final aspects of science they discussed were chaos theory and the anthropic principle. According to the latter everything in the universe needed to be exactly right in order for humankind to survive here—the so-called 'Goldilocks principle'. This has come out of pure science. The only alternative is to postulate many worlds or a multiverse, a complicated, almost science-fiction approach to avoid the common sense conclusion of a Creator/God.

However, as physicists Paul Davies and John Barrow have pointed out, 'We can go no further without revelation.'

It was to this the group turned in their next module, the New Testament. Even the dimensions of M-theory, though an exciting revelation, need to be put alongside the spiritual dimensions when we meet the Holy Spirit. William sometimes talked about Josh being delivered from dark spirits he had picked up from a spiritualist church and from occult involvement and how he was healed in William's room in another college in the name of Jesus and in the power of the Holy Spirit (see Ch. 1). This changed both William and Josh and led William into other areas of healing. As Josh exclaimed, 'I've more proof of the Holy Spirit than of M-theory!' Gifts of healing are the evidence of the Spirit's presence. Heaven is God's presence all around us, as Tom Wright has reminded us.*

The New Testament Module: Meeting the Spirit Again

As they entered the New Testament they left science with the burden of proof of superstrings and dark matter still to come from results at the

* Tom Wright, 'Glossary' in *Acts for Everyone* (London: SPCK, 2008), p. 200.

Large Hadron Collider (LHC) at CERN below Geneva. Working in string theory is almost like a leap of faith—committing one's (research) life to superstrings!

The New Testament frequently became a centre of revelation for the group. Indeed, it was often a commentary which could trigger insight. William was preparing to teach about Paul's letter to the Romans. He recounts, 'As I read about the glory of God's intimate presence, known through the gift of the Spirit, I read about having peace with God "through our Lord Jesus Christ". Chapter 5 goes on to promise hope through suffering, and joy as God has poured out his love into our hearts by the Holy Spirit. I wept in my spirit as I received the peace and love and joy in being united with Jesus in the dimensions of his Spirit. To share the risen life of Jesus was to illuminate my lessons with my own students, "keeping us alert for whatever God will do next"'*

The Dark Dimensions

In one theology session William was speaking about the dark side of spiritualism. Lara asked to see him to talk about her experiences of it. He explained to her that 'contacting the dead' is prohibited in scripture and is a demonic entry point. It also raises false expectations in bereaved people looking for comfort. Using examples such as Josh, who had been freed from the dark spirits he had encountered through such activity, William was able to discourage students from this area. Lara herself had dreams and premonitions of accidents which actually happened, despite her attempts to warn people. 'I didn't want this "seeing-eye" gift,' she admitted. 'You will need to renounce and release these gifts to Jesus,' William explained, 'and he will replace them with *good* gifts.'

New Agers and negative dimensions

Lara later came to tell William that she did now believe in the resurrection of Jesus and his presence today. She had earlier asked if she could initiate an enquirers' group, and she had brought three of her friends to William's lunchtime meetings. An open invitation to talk about spiritual things brought other students during the following weeks.

William was now privately welcoming the Holy Spirit to his classroom each day. Often, as they studied the New Testament they found that

* Romans ch. 5 v. 4 in *The Message,* Copyright © 1993, 1994, 1995, 1996, 2000, 2001, 2002 by Eugene H. Peterson.

they needed to ask for the Holy Spirit's help, as otherwise they found it impenetrable.

William was finding the theology hard going but challenging. He much appreciated the comments of Professor James Dunn, then Lightfoot Professor of Divinity at Durham University, that Paul was 'ransacking his vocabulary' to talk about spiritual things!*

During a discussion on baptism William asked if any of the group had been baptized as an adult. Frankie was happy to talk of her baptism. She described her feelings of going under the water and her sense of the wonder and glorious presence of the Holy Spirit as she emerged. At that very moment, to the group's complete surprise, Stephanie stood up and abruptly declared, 'I'm a witch!' The amazing declaration of the presence of the Holy Spirit had triggered the darkness in Steph! William had a quiet word with her and invited her to come and talk to him after the lesson, which she had before attended only on a trial basis. It was good that she felt able to talk about this. Her mother and grandmother were witches, and she had to make up her own mind when she reached adulthood at the age of twenty-one.

Hostage to the Dark Powers, Negative Dimensions and Negative Energies

The students found it interesting that according to Roger Penrose, Emeritus Professor of Mathematics at Oxford University, the presence of negative energy in physical systems is 'bad news' and 'likely to lead to uncontrollable instabilities'.† This seemed a wonderful analogy for dark, demonic spirits, which, though powerful, can lead to destructive effects on the personality.

Another student who was involved in New Age activities, Elsa, came to the enquirers' group declaring that she was a Wiccan or white witch. She had been to church in the past but had not attended for many years. 'Do you pray against people?' William asked, but she insisted that she prayed only for good things. She found that her prayers worked, but William warned her about the long-term effects and the danger of using her powers when she was angry against someone.

'Could we ask Jesus to send his angels to minister to the people you are praying for?' William asked her. He reminded her that dark spirits were

* *Jesus and the Spirit* (London: SCM Press, 1975), p. 308.

† Roger Penrose, *Cycles of Time* (London: Vintage, 2011), p. 100.

under his authority in Jesus in this college, and that Satan, their master (as she called him), was under the feet of Jesus from the cross. What was she really seeking? William wondered. He told Elsa of her need to repent of seeking occult experiences as a means to find comfort and power. He also gently told her that casting our cares on Jesus and letting go of occult experiences is a process that requires the help of the Holy Spirit. Her friend Simon suggested that they look at an Alpha course, as he was very interested in such a challenge.

The Alpha course

Here Elsa finally realized that you can't earn your way into the Kingdom. It is a free gift by grace from the Spirit's dimension. The only way forward for Elsa and her friends was in forgiveness, knowing the power of Jesus. Elsa was now taking over her own walk in Jesus, and she even renewed church attendance. It was a pleasure for William to find that Stephanie, the former witch, was also now a regular member of the theology group.

Within a year or so the teacher whom William had been called in to replace had given her notice. A new permanent Head of Theology could be appointed; Mark was an excellent choice. William was asked to continue to provide some part-time teaching, but his colleagues were wondering whether the college needed him to work as a Christian counsellor instead. Others told him how valuable they found it to be able to come and talk at the enquirers' group about science and faith, where Elsa had been a catalyst. Simon the atheist became at least a thoughtful agnostic. The principal thanked William for his rescue-teaching at the college—'a real God-send,' she joked. Both in theology lessons and in the enquirers' group, students had become aware of the many dimensions of the Spirit—and in science too!

The Pagan's Tale

At about this time William was asked by the youth pastor from his Christian fellowship to be a 'spiritual mentor' for one of the members of her youth group at another college. 'I'm excited about the possibility,' she wrote to him, 'Andy has a special interest in spiritual warfare.' William was happy to agree to this, so Andy and he met weekly for some months. Andy was one of the students who was into New Age spirituality and paganism in particular. Perhaps he and his friends had been looking for the excitement which few local churches provided. His mother had said he was in need of spiritual counselling, having been badly bullied in his junior school, which resulted in fear and anxiety. She herself was quite concerned whether Andy could be trusted, and he felt that he could never be good enough to be loved.

Andy was happy to take Jesus back to the time of the bullying episodes. He felt Jesus protecting him and asking him to forgive the bullies. William was then able to drive out any spirits of fear and anxiety in the name of Jesus, asking the Holy Spirit to come in to these places. It seemed necessary for Andy to forgive each of his 'friends' by name, and to forgive himself. He was also able to break any words said against him (even by his mother), using the sword of the Spirit. As William commanded any unclean spirits to leave Andy focused on Jesus and felt as if a large snake within him was being hacked to pieces and gradually expelled.

Baptism, and a journey into wholeness

Andy then shared that he was to be baptized on the following Sunday. (Another friend, Mel, was also to be baptized then. Mel had been suffering from severe back pain and found that healing came after taking Jesus back to the time of the injury, forgiving himself and allowing William to ask the Holy Spirit to heal him.) William and Andy agreed that Andy needed the armour of God and that he should keep his guard up, with the healing of a few memories (he found taking Jesus back 'quite scary'). Andy's baptism in the town swimming baths was only the start of his journey into wholeness. A large, heavily built young man with black hair, he burst through the water as 'everything fell away, washed clear'. The following week Andy said that he had had a sense of the Spirit daily since that Sunday—as he put it, 'baptized in water *and* the Spirit in other dimensions'.

Even at this stage Andy needed to renounce and get rid of his old pagan staff, crystals, T-shirt and ring which continued to infect him; this was a gradual renunciation during the next few weeks. He was willing to relinquish rods/wands and a graphic dark hoodie, and later a spirit of addiction and other problems were checked, with the help of two Christian friends. There was much yawning as these came off (yawning is often an unconscious reflex when dark spirits are released) and William and he invited the Holy Spirit to come and bring inner healing from spiritual darkness in the name of Jesus. Andy had been more deeply into New Age paganism than William had realized.

William was given some pictures to share with him. The first one was of a submarine coming to the surface, with the conning tower opening to let out the foul air and draw in the clean, fresh air of the Holy Spirit while the decks were being washed down with a hose. The second was of someone cutting off all the tentacles of a spirit-like octopus, bringing complete freedom.

Andy was now almost completely healed, having given all the big things—his addictions and pagan rites—to the Lord in a deepening awareness of the Father. Even when William entered a formal counselling role Andy would call in to see him. He made good progress with an Open University degree and fed regularly on God's Word, walking in the Spirit. Back at his fellowship he had a new Spiritual Director and was soon actually co-leading an Alpha group himself at a nearby school!

Chapter 9

'A LISTENING EAR'

Counsellor at Cuthbert's College

It had become obvious to the staff that Cuthbert's College needed an official spiritual dimension. Staff members Janet and Andrew both remarked on the need for a Christian counsellor—a friend to chat to, or 'a listening ear', the name William was to give to his office as people came to know and trust him.

Now accepted as an accredited counsellor William worked under supervision as the ACC required. When accepting the voluntary post of counsellor at the college he took a sabbatical from CFM, although he was still under twice-termly supervisions and enjoyed their prayer support. The first request from the hierarchy was that he should be a secular counsellor: 'none of this Jesus stuff,' he was told, half in jest! Using Cognitive Behavioural Therapy as a main pastoral tool helped him to look logically at all the problems that people brought to him. They were able to choose which affected them most and he and they were then able to work on one or two issues over the following weeks. William continued to make himself available for students and teachers to talk to and share any problems. An increasing number came to talk to him as he got to know them through visiting science and art practical sessions, taking occasional lessons in theology or playing the violin with the orchestra or folk group.

Healing Outside the College

Meanwhile, prayer for healing continued in other areas. 'Can I pray for your shoulder and elbow?' William asked Barbara in his church. A week later she came to say how grateful she was, as her shoulder and elbow were much better. John came to give thanks: his hip joint had healed quickly after William had prayed that his operation would go well. Steve had fallen fifteen feet off a ladder and had painfully bruised ribs and kidneys. When William asked him if he would like healing prayer he replied, 'Yes!' Breaking any spirits of trauma in the name of Jesus, William asked the Holy Spirit to bring complete healing to all muscles, sinews, bones and kidneys, and for Steve to receive the peace of Jesus. The next day, somewhat to his surprise, Steve was very much better. However, ministry in the power of the Spirit should not be a spectator sport, and William worked with others whenever possible.

China Prayer Fellowship and Mike the Chemist

Another fellowship William was invited to was a Wednesday China prayer group that met in Mike's house. This was to support Chinese people they knew about, especially the underground church, which meets outside government legislation. On one occasion Mike came home from work with pains in his chest and the back of his neck. The doctor reassured him that he did not have a heart problem but Sylvia, Mike's wife, wanted to cancel that evening's prayer group.

'No!' said Mike. 'Let's get our friends round and pray!', and this was what happened.

That night, after praying for China, the group prayed for deliverance from oppression and stress and the removal of any spirit of anxiety as they asked the Holy Spirit to heal Mike in Jesus' name. After entering the Ezekiel river prayer (the wonderful analogy of entering the river of the Holy Spirit in Ezekiel 47) Mike felt much better, and his recovery was confirmed a day later.

This group was very strong mortar for fellowship. Intercessory prayer was like being helped to climb up one of the towers of the spiritual cathedral—a watchtower—where they felt comfortable with the gifts of the Spirit.

The Chaplain's Log (1)

*Back in Cuthbert's College my work as a counsellor was about to change gear. The College Council invited me to talk about the possibility of being a chaplain to the college. After they approved this, the principal was very happy for me to accept the appointment and invited all the staff to send to me any students who had spiritual problems. Supported by CFM I was able to take on this role, and offered the 'listening ear' of the chaplaincy to all, after each visit 'spring cleaning' the room of the occult attachments of those I saw and filling it with the presence of the Holy Spirit. Staff also came to talk: 'We've needed a chaplain for a long time—welcome!' said Andrew. Others came to share their burdens. 'I feel a lot better having got this off my chest,' said one who felt his gifts weren't being used. I shared the new chaplaincy room with a careers counsellor from CONNEXIONS, a career advisory service 'giving advice and help about jobs, relationships and life', which seemed very suitable. A basic principle behind the service was that both Christian and secular counsellors saw healing as coming 'through unconditional acceptance and forgiveness'.**

The IT expert's tale

'Can I come and talk to you?' asked Mark, once I was settled in my chaplaincy room. He admitted that he often put on a mask when he did his IT work, hiding his New Age experiences and the fact that he heard voices.

'When did it all begin?' I needed to know.

'It all started,' began Mark, 'after a cycle accident about fourteen years ago, when I was nineteen.' Since then he had frequently had a dream that he was in a desert on his own, and he had heard voices and had premonitions/clairvoyance. He admitted that the voices were usually bad, not often very pleasant. That was why he felt the need to put on a mask at work. We agreed that he needed

* Jack Earl in Judy Moore and Campbell Purton, (eds.), *Spirituality and Counselling: Experiential and Theoretical Perspectives* (Ross on Wye: PCCS Books, 2006), p. 294.

to renounce these voices and repent of his occult knowledge of future events.

But first Mark needed to meet Jesus in the power of the Spirit, focusing on Jesus dying on the cross for him, and know that he was indeed special. Mark was able to relate to my story of Josh being healed from voices and dark spirits (see Ch. 1). As he ascended through the dark dimensions in the power of the Holy Spirit Mark started to find the light beaming through everything and driving away the darkness. Accepting Jesus as Lord and Saviour meant that for him there would be no more knowing what was to happen. Mark recognized that this 'knowing for certain' was an occult 'gift', and that speaking against this gift—which seemed to be taking a big risk—was actually speaking out in faith. Mark was soon acting as a mediator and encourager among his family, closing down arguments as each member started learning to forgive as Mark himself had done.

Mark was now able to talk about his faith and his meeting Jesus, and every few weeks he would come to share recent events with me. His family said that in witnessing to others he had become a changed person. Even forgiving himself for the past made everything different; there were no more voices, no more premonitions—all were renounced in the name of Jesus. There would be temptations to return, but this was soon stopped with the words 'Get away in the name of Jesus.'

Spiritual armour

Mark now saw the need to keep on his spiritual armour: the helmet of salvation, the breastplate of righteousness, the belt of truth, 'the sword of the Spirit, which is the Word of God', and to have his feet 'fitted with the readiness that comes from the gospel of peace' and to 'pray in the Spirit'. Mark really appreciated it when I told him that I asked my son if he put on his 'armour' when he and his family were in China and he had replied, 'Dad, I never take it off, not even at night!'*

When witnessing to his parents and brother Mark really felt that he was in a higher dimension of the Spirit. 'Now,' he said, 'there are two

* 'The Armour of God', Ephesians ch. 6 vv. 10–17.

sets of footprints in the sand, not just one set [where Jesus had been carrying him].' Instead of wearing the mask of New Age knowledge of the future he just needed to be with the Holy Spirit. From now on he was a new man every day, right from his early morning times of quiet. He was free from the voices when he went to sleep—instead he asked the Holy Spirit to speak to him—and he had also given up smoking. The dark shadows stopped completely when he received prayer counselling.

Continuing in the Spirit

We were to keep meeting from time to time, even though he left the college for another computer post. Mark still manages meditation in the Spirit before work. He said to me on leaving, 'Your presence produces calm: there is a new balance in the college.' It had been my challenge to create a space for this spiritual dimension to open up—the sense of calm and inner peace which brings healing and wholeness. Mark smiled as he confirmed that he was always better after healing prayer, and that he is now able to help other people who hear voices. Returning from a holiday in Tenerife he said, 'I was in the Spirit all the time, especially at take-off and when landing in turbulence.' He has started to rethink his whole career, 'as a new man, working in constant forgiveness and peace'. 'As for the journey,' he wrote to me, 'it continues. Self-discovery is a wonderful thing; still, trying to achieve a work/life balance is still a challenge.' 'I still wear the armour you helped me to find,' he emailed two years later; 'it's very useful and there's been no darkness for a long time.'

His most recent words to me were, 'I'm in a new dimension, at a good spiritual level'. He was looking forward to a good 'fireside' chat over Christmas. Mark is now married with a baby son.

Spiritually attuned counselling

Students at the college were now becoming aware that there was a spiritual dimension up in the chaplaincy room. They were also used to seeing me around the college, dropping in on practical lessons in sciences and music, and giving an input into theology lessons or playing in the orchestra.

Rebecca talked to me about the whiplash injury affecting her neck. As we broke any spirit of trauma and infirmity she was willing to take Jesus back to the time when she sustained the injury in a car accident, and to forgive herself and the other driver. With Janet, a senior teacher, present I held my hand an inch or two above her neck and shoulder. Rebecca said that she could feel the heat in her neck, though as usual I wasn't making physical contact. I could feel the heat in my own hands as I asked the Holy Spirit to bring healing and peace. Rebecca's neck and shoulders seemed to be completely freed and she went on her way in comfort.

It was a fellow counsellor, Jan, who often shared the chaplaincy room with me. One morning her back was troubling her badly and I asked if I could pray for her. Doubtfully she accepted and was willing to take Jesus back to the time when it started and to forgive herself. We got rid of any trauma and invited the Holy Spirit to come. This proved to be a real healing time for her—a new dimension of healing and counselling. However, a few weeks later, although completely well, she appeared to doubt the spiritual dimension, as sometimes happens. (Susan, a teacher at a different college, was healed in a similar way and also doubted later—although as a curate she admitted she needed to ask for prayer to forgive her vicar first.)

Coping with rejection

Joseph also came to seek my help. An independent student, he reported how he had been bullied at a previous school. He felt as if there was a fog around him, a wall hemming him in. A dark spirit of hatred had come over him as a result of two girls' teasing him. I encouraged him to take Jesus back to these events, but he couldn't forgive. This proved difficult because of other events during his unhappy childhood. Forgiveness is hard; 'it's only possible in Jesus,' I reminded him. 'Father forgive them, they know not what they do' were Jesus' words on the cross. At last Joseph was able to remember the names of the girls and offer them to Jesus. 'He has lifted them both up in his forgiveness,' I said as we asked the Spirit to come and cleanse these dark memories with love, joy and peace. As the Holy Spirit cleansed them like the rooms in a house Joseph saw Jesus break down the doors—and he knew that the Holy Spirit was at work. He was then able to remember the good things of his childhood. His session involved much healing and cleansing.

Too many hurts? The river of the Holy Spirit

There were more hurts from the past which resulted from Joseph's feelings of rejection. 'I don't think I could cope with going back one by one,' he admitted.

It was then, as a way out of dealing separately with these hurts, that I shared the active parable of Ezekiel 47. The river flowing from the temple represents the river of the Holy Spirit. I invited Joseph to enter the river himself and allow the cool water of the Spirit to refresh his feet and the places where he had walked. 'Then, if you will,' I suggested, 'walk further into the stream so that the water covers your knees, where you have prayed; then allow the water to rise further to your waist, cleansing and purifying each part as you go. As the water rises, your chest and heart will be at peace in the healing river of the Holy Spirit. As you allow yourself to go deeper, the words from your mouth and the things you have heard will become purified in the Spirit. Finally, if you are willing to let yourself be submerged, with your head rising above water when necessary, let yourself just enjoy being carried by the spiritual river of love, joy and peace. When you are ready, it's time to swim gently to the bank, where Jesus is waiting with a large bath towel to dry and refresh you in new clean clothes.' Ezekiel 47 ends with the verses, 'Fruit trees of all kinds will grow on both banks of the river. Their leaves will not wither, nor will their fruit fail ... Their fruit will serve for food and their leaves for healing.'

Joseph was filled with wonder and felt empowered by the Holy Spirit to throw off any residual depression—the river taking his thoughts captive as the enclosing 'wall' faded away. He suggested that I use the river parable for my meditation session at a forthcoming retreat for students, and to help another student he knew about who was into the New Age movement.

Now that some of the fog around him had gone he could move forwards and he felt he was being called to tell others about Jesus before it was too late. He spoke to his cousin James at Newcastle University and gave him a copy of my book in manuscript form.

James was amazed to hear Joseph's account of the healing of his memories. Joseph now feels spiritually strong, though he still suffers from the rejection experienced during his childhood. Keeping his 'armour' on has helped, as has using the sword of the Spirit. He realized that healing can take time and that generational ties can be cut at the roots: 'It does work!' he affirmed.

'Are you religious?' James challenged me

As I talked to students around the college I was always made welcome in the art room, where students often revealed some of their deeper feelings—sometimes a darkness within them. One afternoon James asked if he could come and talk to me. (A poster with the heading 'Talk and Share' had been displayed outside the chaplaincy.) He had already sounded out other students 'to see if I was OK to talk to'.

'Are you religious?' was his first question after he knocked on my door.

'No!' I said. I told him that I had been expelled from Sunday school and that my school report said I was 'inclined to fight too readily'! However, I also told him that I did talk about 'spirituality', but this seemed good enough for James. I found that he had a great hunger for reality, for the supernatural dimensions, which was not evident in the churches he had tried. This spirituality is 'less and less related to an institutionalised religion'. (Without the 'supernatural'—or spiritual dimension—the ACC confirmed, counselling can focus on problems and procedures, rather than on Jesus the Deliverer, and even become an alternative religion.)*

James was a lithe, black-haired, earnest student of art; he was also an actor. His mother and grandmother were religious and very influential, but James was not sure about the Spirit. Yet he did accept that events can work out for good if you are following the Holy Spirit. He had joined an acting group in Stockton and had been brilliant in the lead role of Billy Liar. He told me about the director who used the Alexander method, which had led him to 'out of the body' experiences. He had become concerned about this occult development.

---

* Suzette van Ijssel, 'Whose Heaven? The Spiritual Dimension in Humanist Counselling', in *Spirituality and Counselling,* op. cit. p. 268.

However, that seemed enough for him to share that day. 'Thank you so much, it's been a lovely chat,' he said. He had got some worries off his chest and was resolved to treat his future rehearsals with care. (I learned later that James was delighted to be going to Stratford on a studentship.)

Many other students dropped in to talk; some, like Dan, came just to share their doubts and questions.

More Spiritual Warfare: The Chaplain's Log (2)

As I was meeting New Age pagans and witches from time to time I needed to be well versed in the dimensions of spiritual warfare. I was to explore this further during an encounter with Celtic Christianity, as we prepared to take a group of students on a retreat to Holy Island (Lindisfarne). I have no doubt that my earlier boxing training brought dimensions of experience which underpinned this awareness.

The chaplaincy team of myself and two or three other members of staff had long thought that such a retreat would be of great benefit and would show students another dimension. We started our prayer and preparation. As Ged said, 'You're ahead of your time in even starting a chaplaincy in the North-East'; he was pleased that I would still do some extra teaching when required. The principal also showed great joy and pleasure that the college had a chaplain and was 'ahead of the field' in this respect among the other Further Education colleges in the area.

Preparing the way to Holy Island

At the end of one of my last teaching lessons before we went to Holy Island a young student called Hilary came to me. One or two always stayed behind to ask me questions.

'I'm an atheist—would you still pray for me?' she asked hesitantly.

'Of course I will!' I reassured her. 'Jesus never questioned people who came to him.'

Hilary rather shyly told me that she was almost blind in one eye, and had been so from birth. 'Can you pray for my eye?' was her startling demand. The real problem had been that many of her fellow pupils had made fun of her appearance, and this was what really hurt her. However, she was willing to take Jesus back to these hurtful situations (I had just spoken about Jesus in the lesson). When Jesus asked her to do so Hilary was able to forgive each of these pupils who had taunted her. She was overwhelmed by the peace of Jesus when I told her how very much the Father loved her and that she was very precious in his sight. Her whole being changed as she reflected the love and joy of the Father. 'What a lovely person you are!' she exclaimed, but I reassured her that it was not me but the Holy Spirit who was speaking into her heart. Hilary left feeling greatly blessed—but her eye was still the same, as far as I knew. Healing and wholeness had a different meaning for Hilary!

The experiences and teaching at Cuthbert's left William in no doubt about the overlap between science and faith. The students of the college were becoming aware that there was no conflict. The pillars on either side of the cathedral were not only connected to each other, but there were also overarching connections across from science in many dimensions of M-theory to the healing power of the Holy Spirit in the name of Jesus.

Chapter 10

'The New Flatlanders': A Trojan Horse for Jesus

It was back when William was principal of Josh's college (see Ch. 1), and when Josh and his group of students came to ask him questions about how the universe began and the nature of black holes, that William had first encouraged students to enter the story of Flatland and to become Flatlanders themselves. 'It turned me round,' said Thomas, 'it really made me think!' 'You must write a book about this,' the students challenged William; 'students from other colleges need to know about these ideas.' So the concept was developed into a book about science and faith, and was used to challenge many other groups of students, adults, men in prison and even the upper classes.

After receiving twenty or so rejections from publishers William was asked by Pieter and Elria, on one of the publishing boards, if they could be his agents. This was wonderful for William! They soon found a publisher, and while William had been at Cuthbert's his book had been edited, checked and made fit for publication. At last the new book was published (*The New Flatlanders**) and arrived for him to see. It was launched at St John's, Durham, in his own town, and at St Bees, where he first taught chemistry many years before. When asked to give talks all over the North of England William found that the book acted as a 'Trojan horse' for Jesus and the Holy Spirit.

* Eric Middleton, *The New Flatlanders* (Godalming: Highland Books, 2002). Visit www.thenewflatlanders.com/.

The Rugby Player's Tale

So it was that William talked to students in many other colleges and at group meetings. At the 'Probus group' in Durham he was introduced by an old friend, who recommended him to a local sixth form. Giving his usual talk, William described how Josh and others had been healed. 'Can you cure my shoulder?' Angus challenged William in the middle of his talk. This took him unawares, but William asked him to wait behind afterwards, when other theology teachers would still be present. Not knowing what would happen, William was a little apprehensive!

Two other students stayed behind too, and William said some general blessing and healing prayers over them. Angus still stood there, determinedly. He explained that the problem with his shoulder was due to an old rugby injury and that neither doctors nor hospital treatment had made any difference. William felt the Holy Spirit come down on him and he took Angus back to the time of his injury on the rugby field, with Jesus there.

'Will you forgive the person who tackled you too hard?' William asked him.

'Yes,' he agreed.

'And yourself … even God?'

William felt the need to break the spirit of infirmity and of trauma. He held his hand slightly away from Angus's shoulder as he asked the Holy Spirit to bring healing, peace and joy to all the joints, muscles and tendons. Angus felt the heat—although there was no physical contact—as it spread all down his shoulder.

'Flipping heck!' he exclaimed. 'It's working!'

He went off in amazement, happy at the relief and almost jaunty at being set fee from pain. William reminded him of the 'soaking prayer' still needed and to 'gan canny' (go gently). Next day, William could still feel the heat in his hands, which were still sensitive to the Holy Spirit. 'Was this to remind me to keep praying for people and to be ready for more healing when the Spirit leads?' William wondered. From then on, Angus's experience of healing triggered requests for healing from many others following William's talks and lectures on science and spirituality.

It was after he had given about twenty talks across the country that the demands of retreat planning took over William's time and energy.

The Chaplain's Log (3): Holy Island Retreat

Cuthbert's College students were offered the chance to experience a retreat on Holy Island, also known as Lindisfarne, off the coast of Northumberland. Cuthbert's is a secular college and this was an opportunity to stand back ('retreat') from everyday life and reflect on deeper matters. There were to be opportunities for sharing, not least in practical matters: we were self-catering and everyone was expected to be involved.

A spiritual journey: exploring new forms of spirituality

The coastline, sea and rocks, with the untamed landscape and wildlife, certainly affected us. We could understand why the early Celtic monks had regarded this as a 'thin place', where heaven and earth are very close. This was to be a spiritual journey for us all, led by myself, as chaplain, with Mark and Paul, assistant chaplains. Janet, our Buddhist colleague, who was to be a further assistant chaplain, was taken ill at the last moment. Her contribution, 'Spirituality through Art', was taken up by two or three students with pen and sketchbooks. Andrew the artist captured the stones and shadows of the old priory ruins, the isle across the water, and the shapes of the castle on the rock. Our first gathering was on the beach, facing St Cuthbert's Isle with its prominent cross, where Mark inspired us with tales of the Celtic saints Cuthbert, Aidan and Bede. A substantial meal was followed by games, quizzes and good fun on the Thursday evening.

Celtic spirituality for everyone

Andy Raine, one of the original founders of the Northumbria Community on the island, who, with Sister Tessa, was a great support, led that night's meditation. The few who opted out later regretted missing Andy's question-and-answer session which went on into the early hours and followed the teaching of a Celtic praise song, picked up by John on his guitar. This provided space for discussing the spiritual aspects of our lives, helping us to explore our inner spaces, irrespective of 'religion'. In Celtic faith we discovered the strong sense of God being revealed through the natural world and in creativity.

Next morning, after a hearty breakfast (an impromptu grace was said before the meal), everyone joined in, accepting Celtic morning prayer, the wisdom of Mark's musical/visual meditation and Paul's searching self-awareness challenge.

A walk around the beautiful island and historic castle on the outcrop of the Whin Sill caused the group to spread out against the strong wind, talking and sharing. Coffee was followed by the '5p Meditation' ('meditation' was a more comfortable word than 'prayer'), my own adaptation of a key Celtic approach which took the group through the successive stages:*

1. *Pause—rest, stop ... be still ...*
2. *Presence—rest in the presence of the Holy Spirit ... enjoy it, knowing that he is with you now*
3. *Picture—your dreams, visions—or enter the stories and parables, such as Ezekiel 47 (recommended by James), the stilling of the storm (Mark 4), or the Jesus prayer for the healing of memories*
4. *Ponder—the presence of the Spirit, the Spirit of Jesus. 'Remember I am with you always.' This is the reality ...*
5. *Promise—to recall the fact of the presence of Jesus during the day, allowing the Spirit just to touch you and refresh you ...*

When you meet dark spirits, add:

6. *Protection—(a) the Celtic 'Breastplate prayer of Ephesians 6' and (b) the Caim, or encircling prayer, of Psalm 125 v. 2*

We discussed the non-religiosity of Celtic spirituality, knowing that we lacked the language to name and talk about the deeper places or to describe the indescribable.

---

* David Adam, *Living in Two Kingdoms* (London: SPCK, 2007), pp. 12–13.

The Celtic church always took darkness seriously

The Celtic church would not be tolerant of the New Age approaches to creation that make no allowance for the fact that matter can be infected with dark and demonic forces. Using the above prayers for protection they were extremely effective in delivering people and lands from the influences of evil. 'Their spirituality was well-earthed,' Michael Mitton wrote.[*]

Healing of places

It was as principal some years before that I had been asked to bring peace and calm to part of the college—the 'Old House'—which had been left unchanged from earlier days.

With a good Christian colleague we commanded any 'unclean spirits', any spirits of evil which had caused the chilled atmosphere, to depart in the name of Jesus. Then we asked the Holy Spirit to cleanse all the parts of the Old House where these feelings dominated the student council's worries. We asked for a real peace to come upon the place. The atmosphere was restored so that it became a pleasant part of the college from that time. We later learned that an earlier headmaster had committed suicide in that very area, and in the name of Jesus we asked that his soul rest in peace.

On a separate occasion I had felt the Spirit telling me to cleanse my sister's home from dark spirits, although I lived fifty miles away. Using the 'centurion's prayer' of distance healing I went through my sister's house in my spiritual mind's eye, room by room. This was a sort of energetic 'spiritual spring cleaning' which I have found to be effective elsewhere if I am walking in the Spirit myself. Kaye was to tell me a day or two later that a powerful black ball of energy had travelled through her living room at great speed and left through the windows. When we checked the time of this occurrence we found that it was just at the moment when I had been praying the prayer of command in the name of Jesus. Her house was at peace from dark shadows after that day. In each case I had claimed authority over the places.

---

[*] Michael Mitton, *Restoring the Woven Cord* (London: Darton Longman & Todd, 1995), pp. 59, 155.

Dark spirits and 'thin places'

Other areas of great spiritual darkness, such as Glencoe, Rannoch Moor and Chillingham Castle (above the old torture chamber), are beyond what the Spirit will ask me to heal; in such places I must simply keep my armour on! These contrast with the 'thin places' where there have been generations of the prayer dimension, such as Iona, Lindisfarne or Mount Grace Priory. It was interesting that my sister, Kaye, felt the presence of the Spirit in the stillness of the crypt at Durham:

A thousand saints are sleeping
In ethereal splendour, whilst light
Creeps through the Rose Window,
Lifting the spirits of the living.[*]

A lakeside barbecue

Back at Lindisfarne preparations were made, despite the chill wind, for a lakeside barbecue with cheese or ham sandwiches and red wine at the place where we had started our mini-retreat. A tray of charcoal, set alight on the sand, was ready for kippers to be cooked (all had been prepared by Mark in advance).

Gathered in a circle, with St Cuthbert's Isle opposite, we sang Celtic praise songs accompanied by John on his guitar, while the kippers roasted. A final few words from Mark encouraged everyone to share what they had felt on the island and their thoughts on their time together. Ruth readily agreed to read out John's Gospel 21 at this point. We pondered the picture of Jesus by the shore with a fire of burning charcoal, with fish cooking and some bread.

Jesus said to them, 'Come and have breakfast.' None of the disciples dared ask him, 'Who are you?' They knew it was the Lord ... This was now the third time Jesus appeared to his disciples after he was raised from the dead.[†]

... The penny dropped and suddenly the place was indeed a holy island. Then Mark brought us back down to earth: 'This is an agapé

[*] Kaye Norman, 'Durham Cathedral', Jan. 2004 (unpublished).
[†] John ch. 21 vv. 12, 14.

meal,' he said, as we ate the bread, portions of the kippers and drank some wine, looking out across the short stretch of sea to the cross on St Cuthbert's Isle.

There were opportunities for talking, personal guidance and even spiritual mentoring as we waited for the tide to go out, the causeway to clear and the place to be open to pilgrims again.

Looking back

We learned the Celtic way of seeing life as a pilgrimage, of blessing everything that was to happen and of asking for guidance and protection in the Spirit. Back in college verbal feedback revealed conversation taking place at a deeper level without being 'religious'. We had already got to know most of the students through lessons in theology or art, through their reading of my book or because they had been coming for counselling. Many confided that they had begun to know themselves better, know each other better, and know of another dimension which it had not been easy to speak about. The Holy Island experience had opened their eyes; some people were never the same again—myself included! It had not been just a social event, nor just a quiet retreat, but an interaction with a new dimension of a 'spiritual' way of looking at life. Most were inspired by the 'thin place', by the non-religiosity of the Celtic prayers and meditations, and by their courage in sharing what they really felt, especially on that final beach agapé.

Jesus had gone ahead

It had been a memorable two days. 'Living on the edge', we had had only a rough agenda but we knew that Jesus had gone ahead of us (our brief initial prayer), for the Spirit brings order out of chaos. Many were able to experience the transcendence which is at the heart of true spirituality: the awareness of presence, and space for the Spirit to touch us. The Holy Spirit had been amazingly present, particularly in the Celtic prayers as we overlooked St Cuthbert's Isle. Mark wrote to me afterwards, 'We just gave the retreat into God's hands. I was delighted with the response of the students.'

Mark ended his letter, 'All the best for your American tour—it must seem like quite an adventure.' Two weeks later I was on my flight to the USA ...[*]

Chaplain's log ends: 'visiting chaplain' and my shopkeeper dream

It was time for me to think about retiring from the position of counsellor/chaplain while still taking occasional lessons as 'visiting chaplain'. When students like Rachel discovered that these positions were unpaid, fulfilled free of charge, they began to look again at the subject of the presence of the Holy Spirit! So I shared with them a dream from a few years ago.

I was behind the counter in a shop, serving fruit and vegetables. Suddenly, the owner in a white coat stood beside me. 'You are to give these away without charging,' he told me. Seeing that I was quite shocked at this he took me into the store room behind, which was packed full of different produce.

'There's more than sufficient in my reserve for anyone who comes,' he said.

'What are you saying to me, Lord?' was my question, as usual.

'You are not to charge in giving talks about the Flatlanders book,' the reply came.

Occasionally, during my last month as chaplain, I was asked to pray for healing. Alison brought to me Paul, one of her students who had cancer. After this visit Paul found out from his Macmillan nurse that the tumours had started shrinking. I continued to pray with him for healing and blessing, which was to lead to other healings later—my own included!

---

[*] See 'Time Out (2)'.

Time Out (1): My China Challenge: A Vision Of Glory And A New Faith Dimension

With deep prayer cover from CFM, my abiding fellowship of prayer support and love, I was making preparations to fly with KLM to China via Amsterdam. My family had been there for many years; my son Steve had been lecturing in Chinese universities.

The long flight to Beijing was filled with wonderment and anticipation as the plane emerged into the beautiful red dawn of China. Channelled through 'aliens' gates and delayed by lost luggage, at last I was free to meet my son and grandson beyond the barrier. 'Please, Jesus, bring Grandpa to us.' Then eight years old, Michael was sure his prayer would be granted, although it seemed as if Grandpa had missed his flight. Our joyful reunion across 8,000 miles filled us with prayerful thankfulness. As we met friends for breakfast I was asked to pray a blessing on their hopes and challenges as Christians in China. This was a new dimension of prayer: we prayed with our eyes open, as if merely talking over coffee, and I felt the presence of the Father's love for these wonderful people who were teachers and students of Mandarin.

We walked a short section of the 5,000 miles of the Great Wall and this brought startling views: we were watchmen on the wall. Each watchtower was to become a potent symbol for prayer. From Tiananmen Square to the Forbidden City we felt the challenge of living with both yin and yang, the earthly and the heavenly. Flying 2,000 miles west over the Silk Route we were greeted by granddaughter Rachel and Steve's wife, Joy—'Welcome to Xinjiang!'; this was a welcome to many new dimensions. 'I can hardly believe you're really here!' Michael often exclaimed. On long journeys he would often be our intercessor in a natural way. It was an inspiration to me for constantly being with Jesus, for prayer walks round the grounds of Steve's college and entering the Uighur homes of his friends. It was 'out of this world' to meet some of the underground church believers—after dusk in Steve's flat, or down by the river, to talk and pray together, perhaps in Mandarin, in which Steve was by now fluent.

Teaching and healing

Throughout my five-week visit, despite my own weakness and fever, I was constantly aware of the presence of the Spirit. His gifts are always there when needed. So it was to be one evening as Joy, the children and I made our way to the main gate of the campus to go to a nearby restaurant. Steve was to meet us there but it was some time before he appeared, limping very badly. He had twisted his ankle and was in great pain. We decided to take a taxi and get something to eat anyway. Somehow we managed to return home, but then Steve just sank into a chair with his foot up. It was agony for him when Michael simply brushed past him! What should Joy and I do? There were no doctors or hospitals which we could visit in this region of far North-West China. Taking Jesus back to the time it happened Steve was able to forgive himself and be freed from any trauma. We put our hands near his leg, then invited the Holy Spirit to heal every part of his foot and ankle, bringing love and joy and peace. Once I returned to the nearby hotel which was my temporary sleeping quarters I was able, throughout the night, to commit all the family to the Lord and ask for continued healing.

When I returned the next morning Steve seemed to be walking quite well. 'How's the foot?' I ventured to inquire. 'Fine!' he said—there was just a twinge of pain to remind him of the agony he had endured before his ankle was healed. Now I was reassured further—prayer does work.

Under Steve's guidance I was able to talk about Jesus only if his Uighur and Kazak students asked me questions about my faith, or at the 'English Corner'. Steve always had to remain neutral as he was constantly observed in classes by a party member. After flying to Urumqi, the capital of Xinjiang province, I was invited to join a small Christian fellowship. Although a stranger I was trusted because I was Steve's dad, and I became a messenger for some inner healing of problems it was awkward for the sufferers to share with their own friends.

At a Uighur wedding 'banquet' I was challenged to copy the Uighur men's dance. It was a very virile solo dance, controlled by the Master of Ceremonies. He beckoned to the guest of honour to come on the dance floor and show his expertise! I was caught in an unwelcome request but I couldn't refuse, as otherwise I would lose face and Steve and Joy would lose honour. With some acquired panache, a memory of gymnastics, some Scottish country dancing steps (and the power of the Holy Spirit?) I did enough to probably deserve the (sympathetic) applause. The young Americans present were afraid to take the risk of performing in public, while my family sang along happily, beating time on the china bowls with their chopsticks. What a wonderful evening!

On our final visit, to Turpan, the famous oasis on the Silk Route, we entered the infamous Taklimakan Desert ('those who come in may not return'!). It was magical as Michael and Rachel explored the deserted 2,500-year-old city from a different dimension of time. Then it was another mind-blowing experience to shop for silk scarves in the local bazaar. A few weeks later my family waved me off from Urumqi airport on the final stage of my journey via a peaceful night in a Beijing hotel. We flew over the Great Wall of China, easily visible at the end of the day and now outlined in snow.

A room with a view!

However, in the hotel in Beijing I was awoken at 3 a.m.—the Lord wanted to show me something!

I drew back the curtains of my room on the thirteenth floor: there were spare rooms here as no Chinese would accept this 'unlucky' floor number and it was non-smoking! Opposite was a similar skyscraper hotel—but just to its right, level with my eyes, flames and sparks came out of the darkness as if from nowhere … They stopped for a while, but then the flames flickered to life again and seemed to fall downwards. Should I call the fire brigade? Yet where were these flames? In any case, my beginners' Mandarin was not equal to the challenge.

'What are you saying to me, Lord?' The scene was to remain etched on my mind with the prophetic words 'This is my glory coming down on Beijing, on China. Do not let the flame of prayer for China die

down. Do not quench the fire of the Holy Spirit, for Jesus to be known and worshipped in Beijing, in Urumqi and elsewhere …' 'Thank you, Lord': I could only pray in awe and wonder at hearing his voice inside me.

Drawing the curtains I enjoyed a few more hours' sleep, waking to see what had been there. To my daytime eyes there was just another skyscraper being built. There were no lights, and one could only surmise that electricians had been working during the night on its thirteenth floor …

What a beautiful, many-faceted window into the spiritual cathedral which was to become my later vision!

Guardian angels

Following a prayer from Francis McNutt, who was involved in Christian healing ministries, I had been reminded of the presence of our guardian angel in practical matters. Negotiating my final taxi ride I arrived at Beijing airport wondering how I would manage to cope with the decisions ahead in this enormous building filled with Chinese travellers and with signs in Chinese characters. But there, as if waiting personally for me, was a very young man in a bright red uniform. 'Can I help you with your luggage?' he asked. My heart leapt with an assured hope—here was my angel! He placed my heavy luggage on his trolley and proceeded to lead me through the throngs of people to the kiosk to obtain my airport tax. (The need for this had quite slipped my memory.) A 100 yuan note brought 10 yuan change. We then proceeded at a good pace back through the massive crowds into the larger airport space, through swing doors, wending our way another hundred yards and round another corner. There directly in front of us was a baggage point with the notice 'KLM: Amsterdam'!

I was filled with relief and peace; any worries were now behind me … but should I tip an angel? Well, I gave him the 10 yuan note I had received in change! My 'guardian angel' seemed more than satisfied with the outcome.

What a delightful flight back through the many dimensions of space-time: north-west over China, part of Mongolia, then over the frozen tundra of Siberia. Some twelve hours later we came through the airspace of Denmark to land at last in Amsterdam, before the last leg to Newcastle. What stories to share with my dear wife and daughter, waiting at the airport to welcome me home!

Postscript

After my visit my grandson Michael divided the years into 'BG' and 'AG'—'Before Grandpa' and 'After Grandpa'. We exchanged Christmas cards and messages with each of Steve's 120 students, who wrote back with kind words about how they valued Steve's 'firmness and humour'.

We remain watchmen in prayer for the people of China, the Christians serving there (on suitably low wages) and especially the wonderful people of the underground church out in rural China. I now wear two watches: my second watch is my 'prayer watch', worn on my right wrist and set to Xinjiang time, which is seven hours ahead of UK time (as I remembered the non-separability and entanglement of the quantum theory).

Time Out (2): My Mind-Blowing American Adventure

My first book, The New Flatlanders, *originally published in England, was later revised and published by the Templeton Press in America. The door to the States had been opened at a Cambridge conference on science and faith the previous summer. It had been sponsored by Templeton and led to encouragement for me to try them as publishers. My support for Ed Witten's M-theory was challenged by some but I compared his breakthrough with the 150 years it took for Copernicus' theory to be accepted. This proved a useful trigger.*

I had been invited on a two-week book-promoting tour by three or four professors and chaplains, to start just after the Holy Island retreat and supported by the Templeton Foundation. The tour would include open university lectures about my book, talks to student seminars and book signings (often two each day). It would open many windows and memories.

From New Jersey to Oregon

Beginning on the East Coast with Drew University, New Jersey, I was hosted by Professor Leonard Sweet, giving three lectures on the first day. A lecture open to academics and students at lunchtime paved the way for later. 'Science and Spirituality' was followed by 'What Does It Mean to Be Human?' given to a multi-ethnic group. 'That was awesome!' they said as I signed some books. A final meal at an Irish pub with real ale enabled Len and I to share how it had all begun with a cross-Atlantic webcam conversation some months earlier. The table platter was bordered by a Celtic knot—the very symbol of quantum theory 'entanglement' and of the whole visit.

Len wondered if I was up to a 'gig' on the West Coast, where he was Visiting Professor at a university in Portland, Oregon. This would mean travelling to George Fox University and opening further dimensions of opportunity for sharing Jesus.

Psychology and spirituality

When I arrived at Portland, Oregon, the next day I was in time to be a guest speaker on 'Psychology and Spirituality' (I was always introduced as 'Professor'). These lectures were on the interface between psychology and philosophy for postgraduates—new dimensions indeed! Halfway through all my talks I usually changed gear—or 'morphed', as my host Professor Phil Smith described it—into talking about healing in the power of the Spirit. Invariably this led to students remaining behind for prayer for healing—for back problems, anger, voices and so on. After the open lecture in the evening the Celtic knot effect was seen in a wonderful reception given by two professors, Roger and Sue, who had known me many years before when we were Fellows of St John's College, Durham, and who were now based in Oregon. It was at St John's College that my research on Kaluza and the origins of five dimensions had first started. As Bob, one of the professors in Portland, remarked to me,

'All your life has been a preparation for now, perhaps a personal "anthropic principle"!'

Meanwhile, my own dimensions were expanded as my hosts took me to the Pacific coast and also to visit a glass-blowing workshop, besides introducing me to the recherché arts of baseball. Time out with Dwight, my next professor host, included driving up to Mount Hood, snow-topped at 12,000 feet and transcendent in the evening sunset. Later, Professor Paul Anderson and his family and friends welcomed me to their Friends' home group, to hear about the book and to receive prayer requests. As I 'morphed' again they asked, 'Is all that you've told us this evening in the book?' Then they said, 'You must write a second book, "Dimensions of the Spirit".' It was this challenge, coupled with the vision I was given of a cathedral, that resulted in this current book!

More healing of memories

The next day the university chaplain, Sarah, arranged a lunch meeting with some of her friends who were also involved in healing. Jeff and Lucia were from the Spokane group, originators of the 'Healing Rooms' initiatives. Pam, the viola player in my string quartet in Hartlepool, had been telling me how this idea was also spreading in Chester and other places in England. As she had remarked to me, 'You can't ever earn the presence of the Holy Spirit, it is his gift!'

After the final breakfast at George Fox University Dwight had arranged an extended counselling and healing session with one of his postgrads, at her request. Annie and I talked and prayed after breakfast, in the car and at the airport before my flight across to Minnesota. In the name of Jesus she was able to release spirits of anger towards and fear of her dominant religious father, and also to cut off any ungodly soul ties with her violent ex-partner. We asked Jesus to come between her and a sister who always put her down, cutting Annie off from any residual hatred. She also needed to be free from the condemnation of her 'church'. We received a picture of seaweed with long tentacles which had to be cut off, not just with the sword, but with the powerful shears of the Holy Spirit. It only remained to remove any false guilt put on Annie by her church. At her request I prayed over her, asking the Holy Spirit to cleanse every part of her and to bring the love of the heavenly Father and the peace which the Holy Spirit brings in the authority of Jesus. Dwight revealed to me, 'She has been waiting for someone like you for some time.' After much praise and thanks I caught my flight to Minneapolis. Annie and I are still in contact by email, and she has been eagerly awaiting the publication of this book.

On to Minneapolis

Unaware of anyone nearby I was amazed when I felt an 'angel' ready to protect me as I stumbled badly on the airport escalator. I was in Minneapolis (as 'Spike', my nickname at school) to meet an old school friend from Newcastle, now a professor, Ken Reid. He and I had been fellow cricketers, musicians and boxers at Cambridge University and, as it transpired, he was a close friend of my best man at my wedding—a Celtic knot indeed. We enjoyed the talk and book signing at a Wayzata Bookshop and visiting some of the 2,000 lakes and the Minnehaha waterfall of the Hiawatha poem I had learned at junior school. Ken's companion, Roy Kruse, was a member of the Billy Graham Evangelistic Association and after the signing he commented that 'all were touched in many ways'. (I helped with the Cambridge Mission of Billy Graham many years ago.)

Ken was very proud to take me to the nearby St John's University to see the progress on the illuminated Bible. Handwritten and illustrated with scroll pen on parchment, it was the first produced since the invention of the printing press and mirrored the original Lindisfarne Gospels I had seen on Holy Island. There were two large pages open to view: Ezekiel 47 of the river of the Holy Spirit, and Isaiah 61 for inner healing—amazingly apt!

The 'Mars Hill' group

That Friday evening there was another open lecture at the university: 'A Seeker's Guide to the Theory of Everything', 'linking hard science with the spiritual world'. It was arranged by the 'Mars Hill' group of students†—Josiah, Nathan and their group, who were 'committed*

* Text on poster for Minneapolis University lecture, 24 April 2009.

† Their name was inspired by Paul's discussion with philosophers at the Areopagus (Mars Hill) in Athens in Acts 17 (see Acts ch. 17 v. 22 KJV).

to articulate the reality and relevance of their faith to fellow students across the university disciplines' rather than becoming a closed group. My talk, they felt, was ideally suited, and after a very interesting question-and-answer session they invited me for a further fascinating hour's talk chaired by Josiah at a nearby coffee shop (evading an Easter jam/hard rock street band on the way).

They offered me a lift back, but this was a scary experience: running out of petrol late in the evening we had to stop in a vast underground car park, actually part of the overwhelming Mall of America. Josiah left to find petrol and suggested that I find my own way back through the mall. I managed to make my way through the vast circular halls to find a sign saying 'TAXI'! In the pitch black an angel taxi driver found and rescued me, taking me back to my motel and reassuring Josiah that I was home. That night I asked my usual question, 'What are you saying to me, Lord?' The reply was, 'Keep a full tank of the Holy Spirit!' Josiah and I are still in regular and rewarding email contact.

Further companions back from Iraq

A free day allowed me to take the Twin City tour, which included the magnificent cathedral of St Paul's. I was made very welcome on the Sunday by Ken's Baptist church; there was so much enthusiasm about my book that I was overwhelmed by all the American hospitality and kindness. There was just time to say farewell to Ken and his wife before it was time to fly back to Teesside via Amsterdam.

During this trip I was blessed on each of the five flights I took to find companions interested in my book. Some shared with me the nightmares and flashbacks they experienced after their time spent serving in Iraq. We were able to discuss the need for the healing of memories of traumatic events that trigger Post Traumatic Stress Disorder.[*]

On one of my flights I was sat next to a man who worked as a counsellor for returning soldiers who needed help. He was delighted to tell me that recently the authorities of the Red Cross had inserted

[*] I am a member of the Trauma and Abuse Group (TAG), part of the ACC, as recommended by David Woodhouse.

a chapter on 'spirituality' in their counsellors' handbook. Previously, he had been handicapped by a purely secular approach, which, we agreed, was no more effective than sticking plaster.

Dimensions of friendship

A further dimension of friendship retrieved during this visit came when I phoned Foster in a nearby state; he was a friend whom I was unable to visit this time. He had been my host for my lecture to a Pittsburgh University international conference some years before and had thoughtfully also arranged for us to visit the awe-inspiring Niagara Falls. It was after that visit to the States that I first met the Holy Spirit over the Atlantic!

As I flew home I wrote in my Journal, 'I am deeply aware of the Father's love, the joy of the resurrected Jesus and the peace of the many dimensions of the Holy Spirit.' I returned to a family welcome—and to postponed birthday celebrations (I had reached my 74th birthday at the book signing in Minneapolis) which began with the joy of seeing my wife again at Teesside airport and later when the rest of my family came to meet 'Grandpa Dude' back from America!

What amazing, open-minded groups of people I had met as I made new friends and shared about contemporary science in eleven dimensions, the wonderful parable story of Flatland and the gifts of the Holy Spirit in the authority of Jesus! Many students in my audiences had stayed behind to ask for prayer for healing—healing of memories, of voices, and often physical healing. It had been a mind-blowing time in many extra dimensions—it was 'out of this world'!

My Journal entry the day after my return merely records, 'Cut front grass'!

CHAPTER 11

TOOLS AND MORTAR FOR THE 'CATHEDRAL': PRAYER FELLOWSHIP IN THE SPIRIT

Constant fellowship and prayer support had come from William's brothers and sisters in CFM, which he had been invited to join ever since retiring as a principal; they called themselves 'wounded healers'. As accredited counsellors and prayer warriors they make themselves available for any other 'broken stones' which need mending. They are an ecumenical group whose aim is to reach out to the lost and hurting world, as well as to those within churches who need help and want training. They are often given dreams and visions—windows of the Spirit to guide their worship. Other gifts or dimensions of the Spirit are often present, such as tongues and interpretation, prophetic discernment, healing and deliverance.

When members of the group move away they really miss the fellowship, love and prayer. Some have been released from insecurity, perhaps resulting from adoption, into boldness in witnessing. They grew confident in the Holy Spirit, taking risks for Jesus in commanding spirits of anxiety, trauma and infirmity, and even unclean spirits, to go. Others experienced a release from false guilt and have regained self-worth in the love of the Father.

The Blacksmith's Tale

William's friend Edward tells of how he was healed from dark spirits, unforgiveness, and even the spirit of murder when he met the Holy Spirit. A mild, easy-going lad, as the eldest of his family he was given many tasks by his very fierce, brutal and uncaring father. He learnt forgiveness from his mother, who eventually divorced her husband. Edward was made to help kill their hens and pigs (twice-yearly). Despite trying to hide in the shed, he was made to hold each pig's ear while it was slaughtered. He was often beaten by his father and whipped across the face, cutting his ears; this, said Edward, was when the 'spirit of murder' entered him. He would have murdered, given the opportunity!

Edward was removed from school at fourteen to help with jobs on the land, but later he left home to become a stretcher-bearer nurse, foundryman, blacksmith and saddler. Through CFM he met the Holy Spirit and was able to forgive his father and receive Jesus into his heart. Now nearly eighty, Edward is a gardener, attends his local church, sings in the choir with his deep baritone voice and works in the gifts of the Spirit. He has remained one of William's closest friends and he is powerful in intercessory prayer and in the healing of places.

Terry, William's first supervisor, left CFM to run a home for homeless street children in Kasumu, Kenya, devoting his life to their welfare and Christian upbringing. Another Terry became William's new supervisor. A Barnardo's boy who has been healed from many childhood hurts, he is now happily married with a family of his own. He shared with William how he has even been able to forgive Hitler and the German pilots who devastated his family home in London during the war.

Jean's Healing Group

Another close friend from CFM, Jean, told William how she had received much healing, love, help and support from the fellowship group. She was quiet and nervous at first, and she remembers that it was the love she experienced in the group that was totally new. Her experiences led her to become a trained counsellor herself. Jean always paid tribute to Gordon and then to Trevor, his successor as director.

The spirituality of suffering

It is remarkable that, despite her own disability and pain, Jean organizes her own ecumenical healing group. This group has brought generational healing to Deborah, whose health has gradually improved over the years. The saintly Alan has, at the time of writing, terminal cancer, but he looks remarkably well after much prayer and treatment, aided by his caring and loving wife, and is in remission at present. Wholeness is about more than physical healing, Alan told William. 'I'm 80 per cent well, and 100 per cent praising the Lord!' The group were often to find that the experience of inner healing 'while the disease or impairment remains unchecked'* is not uncommon.

Erica's Story

At one prayer support meeting at Jean's house the group were reminded of the visit of Anne Watson to Jean's church some years earlier. Anne's husband, David, vicar of a church in York, was a famed healer and evangelist. Despite worldwide prayer he died of cancer. This had released Anne into her own ministry of prophetic words and healings. During Anne's visit a member of her team was given the name 'Erica'. No one seemed to respond to this until someone remembered that there was a baby called Erica downstairs in the crèche. Anne and another team member went down to find the baby with her mother, Sophie. Anne said how much this baby was loved and cherished by her heavenly Father: 'You are very precious in my sight, and will be a great blessing to others.' Sophie broke down and cried. She already had three children and she hadn't wanted this extra one! Years later, Sophie was able to tell the group how she had come to cherish Erica, who had now grown up to serve the Lord in many ways and was even hoping to become a lay preacher in due course!

Ruth, who has a gift of prophecy and healing, has herself been healed of back pain through the prayers of three or four members of the group. Julia also has a wonderful gift of healing and values prayer. She herself has received healing from ME and for her family. Lynn struggles with ME but says how much the Lord is speaking to her through the illness. Jean's healing group was William's second wonderful fellowship of healing and prayer support.

The Church Alpha Group (Now 'Beta')

The third 'cement mixer of mortar' started with an Alpha group at William's local church. Members of the group found it to be so supportive that it continued long after as a 'Beta' group. This, they said, was their real

* Eric Petrie, *Unleashing the Lion: God's Power in Health and Healing* (London: SPCK, 2000), p. 179.

fellowship, their real church within the large congregation. They studied the scriptures using a commentary with questions, taking it in turns to lead and meeting at one another's houses. William was able to share much of the spiritual advice he had received through his counselling training and they would often feel led to pray for one another's healing.

William asked the members of the Beta group what it had meant to each one. Everyone felt they had gained more understanding of the Holy Spirit and a greater awareness of his presence. They discovered that their Christian faith was not a 'religion' but an experience, a meeting with Jesus in the power of the Spirit. They often prayed that each one would become the person God wanted them to be, according to his loving purpose.

A safe place to share

Beryl, Roger and others had benefited from being able to question things and get a clearer picture of faith and prayer ministry. This was also the case for a vicar's wife, who had felt she had to be loyal to her church's teaching. Julie found (to her husband's surprise) that she was able to conquer her longstanding fear of heights, by always consciously taking the Holy Spirit with her and resting in his peace. In the group every member felt safe to share their more private concerns, their doubts and fears. Marion greatly valued the chance to share her ideas in a small group without fear of ridicule, and to learn persistence in prayer. Kenneth has now been able to forgive more people who have harmed him, such as the Japanese in the last war. However, he finds forgiving someone who burgled his previous house more difficult. He told the group that that was because he could not put a name or a face to the unknown intruder. He has also been helped to find the words to pray, rather than using set prayers. Sophie had come out of a sect (Christian Science) but the aid of doctors and medication produced guilty feelings. While the Lord does use medical treatments for healing (often reinforcing the healing powers of the body itself), within the group Sophie was able to receive healing from false guilt and to enjoy their open fellowship.

The healing room

A number of members of the Beta group have received prayer healing for bad backs, knees or shoulders. In each case they prayed against a spirit of trauma or infirmity, asking the Holy Spirit to heal every nerve, sinew or disc, and they often anointed the sufferer's forehead with olive oil. They were able to minister as a group, having gone through some inner healing themselves. One or two discovered that they needed to say farewell properly to their deceased fathers through Jesus; this brought them great peace and joy. After feeling the awesome presence of the Spirit in the room on one occasion, Vi exclaimed that she was no longer a cynic about the reality of healing! When William invites people to take Jesus back to the time of their injuries for healing and forgiveness, he often says to the group, 'If it feels right, it's for you too!' This presence of the Holy Spirit brings great comfort.

In William's role as leader of another Celtic retreat on Holy Island he talked about healing, only to be asked at the end for healing prayer for backs and shoulders, in small groups. Again, in each case William took Jesus back to the time of the injury, telling the sufferer to forgive whoever had caused it (often themselves) and asking the Holy Spirit to bring healing. As usual, this healing came as William held a hand near the affected part; William would feel heat in his own hand and would see this heat affecting the sufferer's body. Through sharing this the whole group became comfortable with prayer for healing.

Pauline thanked William for talking about binding and loosing in Matthew ch. 18 v. 18. It had given her the confidence to pray a deliverance prayer against dark spirits in an elderly aunt who was having violent nightmares. William had done this himself for his own mother, who had told him about the terrible nightmares she had suffered. He had had to learn to do this. On a later occasion, woken by his wife because of his deep involuntary groaning, he commanded any spirit which was not of Jesus to 'Get out in Jesus' name, never to return!' (This spirit could only have entered while he was unconscious during a recent eye operation.) He often encourages others of all ages to get rid of any nightmares, especially by praying over young children at night.

The Cancer Patient's Tale

It was a new Alpha group started in William's town that attracted Stephen, who asked earnestly for prayer against his terminal cancer. Given only a few weeks to live he was fearful and despondent, and he asked two members of the group if they would pray for him. It was Beryl whom he first approached, and she immediately asked William to join her, her husband Roger and Stephen at her home; the room they used became a real 'healing room' over the next months. Beryl often reminds William of his encouragement 'You have the Holy Spirit, but need to welcome him and be aware of his presence every day … This is because we leak!'.

Stephen's situation needed intensive soaking prayer. They told Stephen to take Jesus back to the start of his cancer, over four years earlier, when he needed to forgive his mother and to be forgiven in Jesus. It turned out that different generations of his family had suffered from cancer, so they took the sword of the Spirit and cut off any generational spirits, especially any spirit of cancer, and asked the Holy Spirit to come into each part of Stephen—blood, tissues, organs. 'How the Father loves you, Steve,' William reassured him. They asked the Holy Spirit to bless the chemotherapy, to heal and renew Stephen, and to work with the doctors. They also suggested that Stephen practised breathing out any spirit of cancer, infirmity or trauma, and breathing in the Holy Spirit.

The Ezekiel prayer of the Holy Spirit again

Then they took Stephen through the Ezekiel river of the Holy Spirit,* letting his feet be covered, then his ankles, knees, thighs and stomach, even his neck, mouth, ears and eyes. They invited the Spirit to bring healing and cleansing to all Stephen's body, and also to the places where he had walked and the things he had seen, heard and spoken. 'Then, if you will,' William said to him, 'allow yourself to be submerged and to swim or float in the spiritual river of life, letting the Holy Spirit wash and refresh you.' When Stephen felt ready they invited him to come out of the river, where Jesus would meet him on the bank, holding out a large white towel. 'How are you?' they asked Stephen. He had felt warm vibrations in his spine and elsewhere in his body, he told them. (Beryl and William had felt this in their hands too.) 'You will need to walk in the Spirit now, and enter the river whenever you like,' William said. Although Stephen gave them profound thanks they immediately replied, 'It wasn't us who did it, it was Jesus!'

'I have met Jesus!'

Stephen walked joyfully downstairs to tell the others in his new Alpha group 'I have met Jesus! It has been like a resurrection!' It was clear that something had happened to him that evening. He had never felt like this before, he exclaimed in wonder, 'although I have been a churchgoer most of my life!' 'Each morning is a resurrection day,' William reassured Stephen; then he told the others, as he often did, 'If it feels right, it's for you too.'

* Ezekiel ch. 47. vv. 1–12.

William slept fitfully that night, with the Holy Spirit continuing to pray soaking prayer within his own spirit.

Three weeks later Stephen reported that the consultant had expected to find more tumours in his body but a scan had showed no further tumours. Stephen put this down to his mentors having the gift of healing. William emphasized that it was due to faith in Jesus and the Holy Spirit.

Dimensions of doubt

Stephen started to have some doubts about whether this healing had really happened. Dark spirits had come in, Stephen said, but under his minister's prayer this heavy load lifted from his shoulders.

At a further session at Stephen's request, with the cancer stabilized, he was able to renounce and repent of Freemasonry. They bound and cast out any spirit of Freemasonry and broke any words said, especially regarding the time he had been given to live, and also cut any further generational ties in this area. 'Praise you, Holy Spirit, for your presence!' they exclaimed. The 'hump on his back' was lifted—he was free! They gave him the Celtic '5p Meditation' (see Ch. 10), adding the sixth 'p' for 'protection with the sword of the Spirit', the Word of God. They felt they should put the sign of the cross on his forehead and on each hand for the gift of healing.

Stephen's testimony: Jesus is real!

It was Stephen and William's mutual friend, Alan, who reminded William that more people resolve the problem of suffering through an encounter with the living Holy Spirit of Jesus than by being offered philosophical or religious answers. When we pray we should first praise God, as Philip Yancey reminds us, for the remarkable agents of healing designed into the body, 'setting in motion the intrinsic powers of healing in a person controlled by God'.*

At this stage Stephen was happy to give his testimony in his church. It was two years since he had been given a week or two to live. Continued visits showed that Stephen was progressing; the aggressive cancer was halted, and as he affirmed, 'Jesus is real! Praise!'

However, a scan a month or two after he gave his testimony showed that the anti-cancer drug he was on was not working and that he must stop taking the other drugs he had been prescribed. Stephen was devastated and felt let down. 'What's the point?' he asked, as the spiritual dark side of stress and fear reappeared. The group were ready to pray further, as

* Philip Yancey, *Prayer: Does It Make a Difference?* (London: Hodder & Stoughton, 2006), p. 318.

William emphasized—and as his friend Alan knew—that while his body might be failing physically, Stephen was improving spiritually. It seems that God may sometimes use our illnesses to bring us to meet Jesus and the Holy Spirit.

Spiritual warfare: the joy of knowing the power of the Spirit!

William was now ready to use the prayer of command with Stephen, another dimension of the Holy Spirit for those who are really close to Jesus: 'In the name of Jesus, I command these cancer cells to wither up and die and these tumours to shrink up and dissolve. From now on, let nothing but normal healthy cells be reproduced in Stephen's body.' In the name of Jesus they commanded that Stephen might be filled with new strength. They invited Stephen to visualize the cancer cells withering away and the tumours shrinking. They also prayed in Jesus' name to bind the spirit of fear, especially fear of cancer and fear of death—to break these with the sword of the Spirit, and for the Holy Spirit to take their place and for Stephen's faith to be strengthened.

Further sacramental healing

At this point they shared Communion of the Lord's Supper, breaking bread and drinking wine after prayer and reading from 1 Corinthians 11. This very powerful sacrament was Jesus' gift to the disciples: 'do this in remembrance of me'. As William reminded Stephen, 'The Father just loves you!' Stephen felt great peace and love as they asked for strength for him to witness to Jesus during his forthcoming sinus operation and to be healed from these dark spirits. They reminded Stephen again to renew the Holy Spirit's presence each day, and they asked for blessing and protection on themselves and their families too.

Next month Stephen phoned: he was very excited—his blood count was much better. The specialist couldn't understand this, Stephen was thrilled to say. 'I feel so much better—I'm really happy at this moment!'

They were greatly encouraged that Stephen had met Jesus, that some healing was taking place (remission, as the specialist would probably say) and that he was free from fear. Stephen continued the prayer of visualizing the disappearance of tumours, and the group agreed to continue to welcome him for further prayer in the Holy Spirit whenever it was needed.

It is now more than two years later, and Stephen and his wife have moved to be nearer their daughter, but he still keeps in touch with William. In September 2011 he wrote, 'I am still going strong under my new consultant. The Lord continues to place his healing hand upon me.' He and his wife are now very happy in a new church.

Prayer in the Fifth Dimension

Prayer has been a multidimensional pillar of the Holy Spirit which was powerfully recommended by David Yonggi Cho, pastor of an enormous and vibrant church in Korea, who talks about praying until you know God's timing. 'You must pray until you have real peace, for peace is like the chief umpire.' He originally referred to prayer as 'the fourth dimension', correcting this later to 'the fifth dimension' in acknowledgement of Einstein's four dimensions of space-time.*

After his experiences with Stephen and prayer in the Holy Spirit, it seemed an appropriate time for William to start writing his next book!

Postscript with Audrey, Marion And Henry: 'Would You Really?'

Soon after this last session with Stephen an elderly lady in William's church to whom William and his wife often gave lifts on Sundays talked about the great pain in her back. She could only walk with difficulty, leaning on her stick. One Sunday William quietly asked her, 'Can I pray for your back, Audrey?'

'Would you really?' she replied in wonderment and joy. 'No one has ever offered to pray for me before!' Audrey was thrilled that, through William, it was clear that the Father knew her and cared for her. 'I'd be only too happy for you to pray for me,' she said and stood up, next to her friend in the pew. 'Is this Reiki?' she wondered.

'No, it's Jesus!' William replied with a smile. He told her to take Jesus back to the time when the sciatica started.

'Jesus is there,' she replied.

'Will you forgive yourself?' (There was no one else to blame!)

'No, I can't!' said Audrey.

'We can go no further unless you will,' he challenged her.

'OK, I have,' she soon responded. William then commanded any spirit of infirmity or trauma to leave Audrey in the name of Jesus. Then he held his hand an inch or two away from the middle of her back, and moved it

* David Yonggi Cho, *The Fourth Dimension,* Vol. 1 (Gainsville, FL: Bridge-Logos, 1983), p. 90; in Volume 2, p. viii, he corrected this, saying, 'I should call it *the fifth dimension*'.

up to her shoulders, neck and even head, asking the Holy Spirit to heal all sinews, discs and nerve connections. Then he commanded her, 'Walk tall!' and he told her that soaking prayer was still needed to keep her back straight and that she needed to sit in upright chairs.

'How are you feeling?' he asked.

'I've got tingling all down my spine!' Audrey replied. 'I feel so much better. I'll be throwing my stick away next week'—and she did!

Meeting in church a week or two later Audrey confessed that she was thrilled to know that someone was actually praying for her. That had started off her recovery, but also she now felt loved, accepted and cared for, and she knew that the Holy Spirit was near. A month after that Audrey was coping well with an additional eye problem and she continued to walk without her stick, except in very cold weather. She was so grateful, she said, for having been healed in just a few minutes and having been set on the right track without her stick. Now that she was looking to Jesus and trusting in the Father she was eager to tell all her family about it.

Three months later a grateful Audrey said she was really very well. Her family relationships had also improved. 'I witness to them about healing and the Holy Spirit. Since you prayed, everything has gone right for me!' Now she is able to travel for longer distances in a motorized wheelchair. She is still praising the Lord, and she now encourages William through *his* illness.

Jolts from the Spirit!

It was while they were in the healing room, praying for Marion's longstanding back problem, that Marion felt two or three sharp blows on her back. Beryl and William assured her that, as usual, they hadn't been touching her; they had just held their hands near her. Marion became much better and slept so well from then on that they could only conclude that the Holy Spirit had been readjusting some discs in her spine!

Taking risks

Although William had been involved in many healings he needed all his courage (as 'Spike' again) to offer to pray for an old friend from his schooldays. Henry, a keen golfer, could not move his neck or shoulders and was unable to play golf one summer. William asked him how it had started: 'Only through turning the wrong way when getting out of bed!' Henry said. He was able to forgive himself (though it had not been his deliberate fault). Commanding any spirit of trauma or infirmity to leave him in the name of Jesus, William asked the Holy Spirit to bring healing, peace and joy to all the muscles and ligaments affected. The result was the disappearance of the pain and discomfort and a release from all his neck and shoulder problems. William warned him about the need for soaking prayer and told him to take things quietly. Henry's wife phoned a day or two later to say that he was free from pain and already back to playing golf! This healing was confirmed a few months later.

The Stonemason's Memoirs

William's grandfather was a stonemason who either used a mallet and chisel for carving or more probably used mortar to lay stones in their correct position in a building. These were originally known as 'setters' but today go by the name of 'fixers'. The stonemason's tools were used to cut, split and dress blocks of stone for walls and pillars, usually under the supervision of a master mason. The smooth-dressed surface of fine sandstone, freestone, in Durham allows it to be cut and worked freely. When building the cathedral, chips of stone were removed from the basic shapes, finer tools being employed at each stage, and passed on to the 'fixer' mason, who set the finished stone precisely in place.

Lime mortar was used because it is a soft, porous and malleable material and helps the cathedral to move and even crack without major damage. 'It acts like a cushion to every jointed stone and helps the structure breathe,' wrote Thomas Maude in *Guided by a Stone-Mason.*[*] This seems a very striking analogy for these chapters on the tools needed for healing and on the mortar of fellowship.

[*] Thomas Maude, *Guided by a Stone-Mason: Cathedrals, Abbeys and Churches of Britain Unveiled* (London: I. B. Taurus & Co; repr. 2009), p. 88.

Chapter 12

THE SPIRITUAL CATHEDRAL: A VISUAL METAPHOR

The vision of a spiritual cathedral which William was given has become clearer through this account of healings. It seemed to bear a resemblance to his nearby cathedral at Durham. The baptismal font at the entrance stands for the entry points to the spiritual dimensions—encountering Jesus and the Holy Spirit. The length of the cathedral stretches ahead, the parallel rows of pillars appearing to converge towards the beautiful east window, the sanctuary shrine of St Cuthbert, upon whose Celtic faith the foundations were originally laid, and the high altar of the cross of Jesus. William is reminded that the word for the main body of a cathedral, the nave, comes from the Latin meaning 'a ship'—a means of transport for his spiritual voyage to the many dimensions of healing and also of exploring contemporary science. At the heart of the nave is the horizontal crossover, the 'footprint' symbol of the cross and resurrection. Here there was once a screen separating the everyday from the holy, the spiritual from the secular. Now the two are interwoven, reminding William of Paul Davies' saying that 'we can go no further without revelation'. The great tower provides a magnificent view—it is a tower of intercessory prayer.

The great 'piers' or pillars move eastwards, reinforcing the cathedral as a sequence of thresholds. The pillars are the stepping stones of William's experiences. On one side the pillars seem to be the great discoveries of science, unified by arches connecting each one to the next. They begin with the separate small pillars of electricity and magnetism which are united by an arch to the main pillar of electromagnetism. This appeared to William as light and radio waves, linked across the nave by Jesus' saying, 'I am the light of the world' (John ch. 8 v. 12).

It doesn't take an Einstein to understand the modern physics of curved space. The strong pillar of Einstein's relativity towers above, with its spin-offs of the singularities of the Big Bang and black holes, perhaps entry points to other dimensions.

William's explorations in science took him to the next great pillar of the twentieth century, quantum theory. Although the basis of much of the technology we use today, its explanations defy common sense. The mystery and paradoxes are seen in observer-centred reality (where Jesus must be the real Observer), and entanglement, or Einstein's 'spooky action at a distance'. This links with the Celtic knot idea of the non-separability of the Father's love—and also the 'spooky' action of the power of the Holy Spirit across every dimension.

The Critical Key in Five Dimensions

The key arch linking apparently disparate pillars of science started with Theodor Kaluza. Almost a hundred years ago his great discovery was to go beyond the three dimensions of space and time: 'Let's try five dimensions!' It proved to be the key to unifying Einstein's awesome relativity with the mysteries and paradoxes of quantum theory. He was ahead of his time, yet 'It worked!' Discovering that the atom was mainly empty space, its centre was found to consist of quarks. Although never seen, like the even tinier strings, they are now accepted as reality—'because they work!' So the enormous pillars morphed into multidimensional columns of ten or eleven dimensions, and then into the widely accepted theory of everything, M-theory in eleven dimensions. These science pillars revealed other languages for talking, first the anthropic principle, in which everything in the universe is seen to be fine-tuned so that man can survive here. A further language was found in chaos theory, from which William and his students saw that Jesus was the real Fractal, the hidden image of the Creator God within chaos itself.

These discoveries provided wonderful analogies with the pillars down the other side of the cathedral. The wonder and awe of the universe was echoed in the spiritual 'big bang' of Josh's healing from dark spirits, revealing Jesus to be the real singularity, the real entry point to the dimensions of the Spirit. Contemporary science was linked with the work of the Holy Spirit in the amazing testimonies and pillars of healing and wholeness.

What had begun as just listening became praying into God's dimensions all around. Inner healing, healing of memories and generational healing often led to physical healing. William and his colleagues crucially discovered that forgiveness is healing, the most powerful dimension in Jesus through the power of the Spirit.

The Celtic sense of the Spirit's presence in everything took them to the mind-blowing idea that reality is more than three dimensions. They could see that the magnificent arches connecting the two rows of pillars across the nave were bridging the divide. They combined the extra dimensions needed in physics with the powerful extra dimensions of the Spirit.

The jigsaw came together in finding a language to even think about such things. It was the wonderful parable story of Flatland that really provided a focus for students of all ages. This apparently secular story brought down to earth the concept of eleven extra dimensions and revealed that the sphere who entered Flatland himself revealed Jesus to us. 'Anyone who has seen me has seen the Father' (John ch. 14 v. 9). We are only a three-dimensional shadow of a higher dimensional reality.

Analogies leapt to mind of the tools of prayer ministry to shape and smooth the stones, and also the binding mortar of prayer and fellowship. They found they were unable to go further without revelation. For William, this came with his harbour vision, his experience on Brighton beach or his shopkeeper's dream and many other times when the Spirit spoke. 'Do you pray into other dimensions?' asked Ros, a friend in William's prayer group. 'Of course,' William replied. 'Even saying the Lord's Prayer is praying into other dimensions all around us—to know that in prayer I can cross the boundaries of space and time and be with my brothers and sisters in the Spirit wherever they are.' As Brother Ramon wrote, 'One believer's prayers send out vibrations and reverberations that increase the power of the divine Love in the cosmos.'*

* Brother Ramon, *A Hidden Fire: Exploring the Deeper Reaches of Prayer* (Basingstoke: Marshall Pickering, 1987), pp. 155–156.

An Artist's Spiritual Dimensions: Lois the Poet

'It all makes sense now, it all hangs together,' wrote William's friend Lois, poet, painter and lover of music. 'I can see the cathedral interior quite clearly, and the pillars of recent science and healings. Also, the tools to shape and smooth the stones involving inner healing, etc.; and I can sense the mortar of prayer and fellowship groups. I love the thought of the foundations being built on Celtic Christianity from Holy Island. M-theory I know is "more than this world dreams of". We are only a three-dimensional shadow of a higher dimensional reality! But have you any sound, is there colour as well as light through the windows? Whatever the terms, the vision is very vivid!'

William wrote to reassure her of a minstrel's gallery, spaces for music, Celtic folk music, choirs and so on. 'Yes, the dimensions of colour are in the Rose Window, the windows commemorating Celtic saints—Cuthbert and Bede from Holy Island.'

Making Connections

These experiences were to trigger a wake-up call for a new world view in many dimensions. In this book we have seen that:

(1) The healing of hurtful memories, trauma or voices is a reality—as is also physical healing, and the healing of words and places.

(2) We need not be afraid of contemporary science—a new world view, including the extra dimensions needed in the physics of M-theory, as we wait for strengthening evidence from the LHC beneath Geneva. It seems to make sense for physics today—and for the theology of the resurrection and the healing power of the Holy Spirit.

(3) It is indeed a sacramental universe where, in the end times, the hidden dimensions may be unfurled, and the three dimensions of our everyday experience be curled up in their turn ...

(4) Stephen Hawking's words have come true: 'if we do find the "Theory of Everything", we will know the Mind of God.'* The Holy Spirit speaks to us today through people, in the Word of scripture and in events—three-dimensional shadows become indeed of a higher dimensional reality!

Stepping stones

It was in humble wonder,' William writes, 'that I have traced my own stepping stones from science to spirituality, which became pillars as I became

* Stephen Hawking, op. cit. p. 175.

more conscious of the Holy Spirit's presence. Available to all who welcome him, without any pride or unforgiving spirit, he was to be the spiritual power behind all the testimonies, all the stories, I was invited to share.'

The Celtic Foundation

The stories in this book are both spiritual Canterbury tales *and* the building blocks of the spiritual cathedral that we are working towards. It is time to return to the Celtic foundation of William's spiritual adventures, remembering the 5 ps: Pausing to be aware of the Spirit's Presence, William often ends his day with a Picture to Ponder and a Promise to Recall, from *Celtic Daily Prayer*:

Calm me, O Lord, as You stilled the storm,
Still me, O Lord, keep me from harm.
Let all the tumult within me cease,
Enfold me, Lord Jesus, in Your peace.*

Postscript from William

It is often our experience that it is not so easy to pray for our own healing. As we end the account of these adventures in the Spirit the thoughts and prayers of all my friends are for the surgeons, nurses and helpers as they bring healing for my own cancer. Successful operations for bowel cancer have brought me great peace, even as I now end a lengthy treatment with ongoing chemotherapy …

It has been wonderful to know the heartfelt support and prayers from many of those mentioned in this book: 'You are in my prayers and on the circuit prayer list—take care!' wrote Pam, the retired senior tutor. 'I'm deeply saddened to hear your news, such a shock; I don't mind telling you it almost brought me to tears. I still wear the armour you helped me to find—against darkness … Praying for a speedy recovery.' 'Do keep my name in your book. After all, you did help me on so many levels. I've never forgotten our chats, nor will I,' said Mark, the IT expert now working in music and media.

* 'Wednesday Compline', in *Celtic Daily Prayer: Inspirational Prayer and Readings from the Northumbria Community* (London: Harper Collins, 2000), p. 38.

Often people like Sue ask me how the cancer is progressing, and I take the opportunity to tell them about the healing of Martin or Josh. They are then ready to believe—because they trust me to tell them how it is! May this book challenge my readers also ...

Lightning Source UK Ltd.
Milton Keynes UK
UKOW032139080213

206040UK00010B/373/P